de Gruyter Expositions in Mathematics 37

de Gruyter Expositions in Mathematics

1 The Analytical and Topological Theory of Semigroups, *K. H. Hofmann, J. D. Lawson, J. S. Pym* (Eds.)
2 Combinatorial Homotopy and 4-Dimensional Complexes, *H. J. Baues*
3 The Stefan Problem, *A. M. Meirmanov*
4 Finite Soluble Groups, *K. Doerk, T. O. Hawkes*
5 The Riemann Zeta-Function, *A. A. Karatsuba, S. M. Voronin*
6 Contact Geometry and Linear Differential Equations, *V. E. Nazaikinskii, V. E. Shatalov, B. Yu. Sternin*
7 Infinite Dimensional Lie Superalgebras, *Yu. A. Bahturin, A. A. Mikhalev, V. M. Petrogradsky, M. V. Zaicev*
8 Nilpotent Groups and their Automorphisms, *E. I. Khukhro*
9 Invariant Distances and Metrics in Complex Analysis, *M. Jarnicki, P. Pflug*
10 The Link Invariants of the Chern-Simons Field Theory, *E. Guadagnini*
11 Global Affine Differential Geometry of Hypersurfaces, *A.-M. Li, U. Simon, G. Zhao*
12 Moduli Spaces of Abelian Surfaces: Compactification, Degenerations, and Theta Functions, *K. Hulek, C. Kahn, S. H. Weintraub*
13 Elliptic Problems in Domains with Piecewise Smooth Boundaries, *S. A. Nazarov, B. A. Plamenevsky*
14 Subgroup Lattices of Groups, *R. Schmidt*
15 Orthogonal Decompositions and Integral Lattices, *A. I. Kostrikin, P. H. Tiep*
16 The Adjunction Theory of Complex Projective Varieties, *M. C. Beltrametti, A. J. Sommese*
17 The Restricted 3-Body Problem: Plane Periodic Orbits, *A. D. Bruno*
18 Unitary Representation Theory of Exponential Lie Groups, *H. Leptin, J. Ludwig*
19 Blow-up in Quasilinear Parabolic Equations, *A. A. Samarskii, V. A. Galaktionov, S. P. Kurdyumov, A. P. Mikhailov*
20 Semigroups in Algebra, Geometry and Analysis, *K. H. Hofmann, J. D. Lawson, E. B. Vinberg* (Eds.)
21 Compact Projective Planes, *H. Salzmann, D. Betten, T. Grundhöfer, H. Hähl, R. Löwen, M. Stroppel*
22 An Introduction to Lorentz Surfaces, *T. Weinstein*
23 Lectures in Real Geometry, *F. Broglia* (Ed.)
24 Evolution Equations and Lagrangian Coordinates, *A. M. Meirmanov, V. V. Pukhnachov, S. I. Shmarev*
25 Character Theory of Finite Groups, *B. Huppert*
26 Positivity in Lie Theory: Open Problems, *J. Hilgert, J. D. Lawson, K.-H. Neeb, E. B. Vinberg* (Eds.)
27 Algebra in the Stone-Čech Compactification, *N. Hindman, D. Strauss*
28 Holomorphy and Convexity in Lie Theory, *K.-H. Neeb*
29 Monoids, Acts and Categories, *M. Kilp, U. Knauer, A. V. Mikhalev*
30 Relative Homological Algebra, *Edgar E. Enochs, Overtoun M. G. Jenda*
31 Nonlinear Wave Equations Perturbed by Viscous Terms, *Viktor P. Maslov, Petr P. Mosolov*
32 Conformal Geometry of Discrete Groups and Manifolds, *Boris N. Apanasov*
33 Compositions of Quadratic Forms, *Daniel B. Shapiro*
34 Extension of Holomorphic Functions, *Marek Jarnicki, Peter Pflug*
35 Loops in Group Theory and Lie Theory, *Péter T. Nagy, Karl Strambach*
36 Automatic Sequences, *Friedrich von Haeseler*

Error Calculus for Finance and Physics:

The Language of Dirichlet Forms

by

Nicolas Bouleau

Walter de Gruyter · Berlin · New York

Author

Nicolas Bouleau
École Nationale des Ponts et Chaussées
6 avenue Blaise Pascal
77455 Marne-La-Vallée cedex 2
France
e-mail: bouleau@enpc.fr

Mathematics Subject Classification 2000:
65-02; 65Cxx, 91B28, 65Z05, 31C25, 60H07, 49Q12, 60J65, 31-02, 65G99, 60U20, 60H35, 47D07, 82B31, 37M25

Key words:
error, sensitivity, Dirichlet form, Malliavin calculus, bias, Monte Carlo, Wiener space, Poisson space, finance, pricing, portfolio, hedging, oscillator.

∞ Printed on acid-free paper which falls within the guidelines of the ANSI to ensure permanence and durability.

Library of Congress – Cataloging-in-Publication Data

Bouleau, Nicolas.
Error calculus for finance and physics : the language of Dirichlet forms / by Nicolas Bouleau.
p. cm – (De Gruyter expositions in mathematics ; 37)
Includes bibliographical references and index.
ISBN 3-11-018036-7 (alk. paper)
1. Error analysis (Mathematics) 2. Dirichlet forms. 3. Random variables. I. Title. II. Series.
QA275.B68 2003
511′.43–dc22 2003062668

ISBN 3-11-018036-7

Bibliographic information published by Die Deutsche Bibliothek

Die Deutsche Bibliothek lists this publication in the Deutsche Nationalbibliografie; detailed bibliographic data is available in the Internet at <http://dnb.ddb.de>.

Typesetting using the authors' TEX files: I. Zimmermann, Freiburg.
Printing and binding: Hubert & Co. GmbH & Co. KG, Göttingen.
Cover design: Thomas Bonnie, Hamburg.

Preface

To Gustave Choquet

Our primary objective herein is not to determine how approximate calculations introduce errors into situations with accurate hypotheses, but instead to study how rigorous calculations transmit errors due to inaccurate parameters or hypotheses. Unlike quantities represented by entire numbers, the continuous quantities generated from physics, economics or engineering sciences, as represented by one or several real numbers, are compromised by errors. The choice of a relevant mathematical language for speaking about errors and their propagation is an old topic and one that has incited a large variety of works. Without retracing the whole history of these investigations, we can draw the main lines of the present inquiry.

The first approach is to represent the errors as random variables. This simple idea offers the great advantage of using only the language of probability theory, whose power has now been proved in many fields. This approach allows considering error biases and correlations and applying statistical tools to guess the laws followed by errors. Yet this approach also presents some drawbacks. First, the description is too rich, for the error on a scalar quantity needs to be described by knowledge of a probability law, i.e. in the case of a density, knowledge of an arbitrary function (and joint laws with the other random quantities of the model). By definition however, errors are poorly known and the probability measure of an error is very seldom known. Moreover, in practical cases when using this method, engineers represent errors by means of Gaussian random variables, which means describing them by only their bias and variance. This way has the unavoidable disadvantage of being incompatible with nonlinear calculations. Secondly, this approach makes the study of error transmission extremely complex in practice since determining images of probability measures is theoretically obvious, but practically difficult.

The second approach is to represent errors as infinitely small quantities. This of course does not prevent errors from being more or less significant and from being compared in size. The errors are actually small but not infinitely small; this approach therefore is an approximate representation, yet does present the very significant advantage of enabling errors to be calculated thanks to differential calculus which is a very efficient tool in both the finite dimension and infinite dimension with derivatives in the sense of Fréchet or Gâteaux.

If we apply classical differential calculus, i.e. formulae of the type

$$dF(x, y) = F_1'(x, y)\, dx + F_2'(x, y)\, dy$$

we have lost all of the random character of the errors; correlation of errors no longer has any meaning. Furthermore, by nonlinear mapping, the first-order differential calculus applies: typically if $x = \varphi(s,t)$ and $y = \psi(s,t)$, then $dx = \varphi_1' \, ds + \varphi_2' \, dt$ and $dy = \psi_1' \, ds + \psi_2' \, dt$, and

$$dF\big(\varphi(s,t), \psi(s,t)\big) = \big(F_1'\varphi_1' + F_2'\psi_1'\big)\, ds + \big(F_1'\varphi_2' + F_2'\psi_2'\big)\, dt.$$

In the case of Brownian motion however and, more generally, of continuous semi-martingales, *Itô calculus* displays a second-order differential calculus. Similarly, it is indeed simple to see that error biases (see Chapter I, Section 1) involve *second derivatives* in their transmission by nonlinear functions.

The objective of this book is to display that errors may be thought of as *germs of Itô processes*. We propose, for this purpose, introducing the language of Dirichlet forms for its tremendous mathematical advantages, as will be explained in this book. In particular, this language allows error calculus for infinite dimensional models, as most often appear in physics or in stochastic analysis.

<table>
<tr><td rowspan="2">Deterministic approaches</td><td colspan="2">Deterministic sensitivity analysis:
derivation with respect to the parameters of the model</td><td rowspan="2">Interval calculus</td></tr>
<tr></tr>
<tr><td rowspan="2">Probabilistic approaches</td><td colspan="2">Error calculus using Dirichlet forms</td><td rowspan="2">Probability theory</td></tr>
<tr><td>first order calculus only dealing with variances</td><td>second order calculus with variances and biases</td></tr>
<tr><td></td><td colspan="2">Infinitesimal errors</td><td>Finite errors</td></tr>
</table>

The approach we adopt herein is therefore intermediate: the errors are infinitely small, but their calculus does not obey classical differential calculus and involves the first and second derivatives. Although infinitely small, the errors have biases and variances (and covariances). This aspect will be intuitively explained in Chapter I.

The above table displays the various approaches for error calculations. It will be commented on in Chapter V, Section 1.2. Among the advantages of Dirichlet forms (which actually limit Itô processes to symmetric Markovian processes) let us emphasize here their *closed character* (cf. Chapters II and III). This feature plays a similar role in this theory to that of σ-additivity in probability theory. It yields a powerful extension tool in any situation where the mathematical objects through which we compute the errors are only known *as limit* of simpler objects (finite-dimensional objects).

This text stems from a postgraduate course taught at the Paris 6 and Paris 1 Universities and supposes as prerequisite a preliminary training in probability theory. Textbook references are given in the bibliography at the end of each chapter.

Acknowledgements. I express my gratitude to mathematicians, physicists and finance practitioners who have reacted to versions of the manuscript or to lectures on error calculus by fruitful comments and discussions. Namely Francis Hirsch, Paul Malliavin, Gabriel Mokobodzki, Süleyman Üstünel, Dominique Lépingle, Jean-Michel Lasry, Arnaud Pecker, Guillaume Bernis, Monique Jeanblanc-Picqué, Denis Talay, Monique Pontier, Nicole El Karoui, Jean-François Delmas, Christophe Chorro, François Chevoir and Michel Bauer. My students have also to be thanked for their surprise reactions and questions. I must confess that during the last years of elaboration of the text, the most useful discussions occurred from people, colleagues and students, who had difficulties understanding the new language. This apparent paradox is due to the fact that the matter of the book is emerging and did not yet reach a definitive form. For the same reason is the reader asked to forgive the remaining obscurities.

Paris, October 2003 *Nicolas Bouleau*

Contents

Chapter I

Intuitive introduction to error structures

Learning a theory is made easier thanks to previous practical training, e.g. probability theory is usually taught by familiarizing the student with the intuitive meaning of random variables, independence and expectation without emphasizing the mathematical difficulties. We will pursue the same course in this chapter: managing errors without strict adherence to symbolic rigor (which will be provided subsequently).

1 Error magnitude

Let us consider a quantity x with a small centered error εY, on which a nonlinear regular function f acts. Initially we thus have a random variable $x + \varepsilon Y$ with no bias (centered at the true value x) and a variance of $\varepsilon^2\sigma_Y^2$: $\text{bias}_0 = 0$, $\text{variance}_0 = \varepsilon^2\sigma_Y^2$.

Once the function f has been applied, use of Taylor's formula shows that the error is no longer centered and the bias has the same order of magnitude as the variance.

Let us suppose that f is of class C^3 with bounded derivatives and with Y being bounded:

$$\begin{aligned}
f(x + \varepsilon Y) &= f(x) + \varepsilon Y f'(x) + \frac{\varepsilon^2 Y^2}{2} f''(x) + \varepsilon^3 O(1) \\
\text{bias}_1 &= \mathbb{E}[f(x + \varepsilon Y) - f(x)] = \frac{\varepsilon^2 \sigma_Y^2}{2} f''(x) + \varepsilon^3 O(1) \\
\text{variance}_1 &= \mathbb{E}\big[(f(x + \varepsilon Y) - f(x))^2\big] = \varepsilon^2 \sigma_Y^2 f'^2(x) + \varepsilon^3 O(1).
\end{aligned}$$

Remark. After application of the non-linear function f some ambiguity remains in the definition of the error variancc. If we consider this to be the mean of the squared deviations from the true value, we obtain what was previously written:

$$\mathbb{E}[(f(x + \varepsilon Y) - f(x))^2];$$

however, since the bias no longer vanishes, we may also consider the variance to be the mean of the squared deviations from the mean value, i.e.,

$$\mathbb{E}[(f(x + \varepsilon Y) - \mathbb{E}[f(x + \varepsilon Y)])^2].$$

This point proves irrelevant since the difference between these two expressions is

$$\big(\mathbb{E}[(f(x+\varepsilon Y)] - f(x)\big)^2 = \varepsilon^4 O(1)$$

which is negligible.

If we proceed with another nonlinear regular function g, it can be observed that the bias and variance of the error display the same order of magnitude and we obtain a transport formula for small errors:

$$(*) \qquad \begin{cases} \text{bias}_2 = \text{bias}_1 g'(f(x)) + \frac{1}{2}\text{variance}_1 g''(f(x)) + \varepsilon^3 O(1) \\ \\ \text{variance}_2 = \text{variance}_1 \cdot g'^2(f(x)) + \varepsilon^3 O(1). \end{cases}$$

A similar relation could easily be obtained for applications from $\mathbb{R}^p$ into $\mathbb{R}^q$.

Formula $(*)$ deserves additional comment. If our interest is limited to the main term in the expansion of error biases and variances, *the calculus on the biases is of the second order and involves the variances. Instead, the calculus on the variances is of the first order and does not involve biases*. Surprisingly, calculus on the second-order moments of errors is simpler to perform than that on the first-order moments.

This remark is fundamental. Error calculus on variances is necessarily the first step in an analysis of error propagation based on differential methods. This statement explains why, during this entire course, emphasis is placed firstly on error variances and secondly on error biases.

2 Description of small errors by their biases and variances

We suppose herein the usual notion of conditional expectation being known (see the references at the end of the chapter for pertinent textbooks).

Let us now recall some notation. If X and Y are random variables, $\mathbb{E}[X \mid Y]$ is the same as $\mathbb{E}[X \mid \sigma(Y)]$, the conditional expectation of X given the σ-field generated by Y. In usual spaces, there exists a function φ, unique up to $\mathbb{P}_Y$-almost sure equality, where $\mathbb{P}_Y$ is the law of Y, such that

$$\mathbb{E}[X \mid Y] = \varphi(Y).$$

The conventional notation $\mathbb{E}[X \mid Y = y]$ means $\varphi(y)$, which is defined only for $\mathbb{P}_Y$-almost-every y.

We will similarly use the *conditional variance*:

$$\text{var}[X \mid Y] = \mathbb{E}\big[(X - \mathbb{E}[X \mid Y])^2 \mid Y\big] = \mathbb{E}[X^2 \mid Y] - (\mathbb{E}[X \mid Y])^2.$$

There exists ψ such that $\text{var}[X \mid Y] = \psi(Y)$ and $\text{var}[X \mid Y = y]$ means $\psi(y)$, which is defined for $\mathbb{P}_Y$-almost-every y.

2.1. Suppose that the assessment of pollution in a river involves the concentration C of some pollutant, with the quantity C being random and able to be measured by an experimental device whose result exhibits an error ΔC. The random variable ΔC is generally correlated with C (for higher river pollution levels, the device becomes dirtier and fuzzier). The classical probabilistic approach requires the *joint law* of the pair $(C, \Delta C)$ in order to model the experiment, or equivalently the law of C and the conditional law of ΔC given C.

For pragmatic purposes, we now adopt the three following assumptions:

A1. We consider that the conditional law of ΔC given C provides excessive information and is practically unattainable. We suppose that only the conditional variance $\mathrm{var}[\Delta C \mid C]$ is known and (if possible) the bias $\mathbb{E}[\Delta C \mid C]$.

A2. We suppose that the errors are small. In other words, the simplifications typically performed by physicists and engineers when quantities are small are allowed herein.

A3. We assume the biases $\mathbb{E}[\Delta C \mid C]$ and variances $\mathrm{var}[\Delta C \mid C]$ of the errors to be of the *same* order of magnitude.

With these hypotheses, is it possible to compute the variance and bias of the error on a function of C, say $f(C)$?

Let us remark that by applying A3 and A2, $(\mathbb{E}[\Delta C \mid C])^2$ is negligible compared with $\mathbb{E}[\Delta C \mid C]$ or $\mathrm{var}[\Delta C \mid C]$, hence we can write

$$\mathrm{var}[\Delta C \mid C] = \mathbb{E}[(\Delta C)^2 \mid C],$$

and from

$$\Delta(f \circ C) = f' \circ C \cdot \Delta C + \frac{1}{2} f'' \circ C \cdot (\Delta C)^2 + \text{negligible terms}$$

we obtain, using the definition of the conditional variance,

$$(1) \qquad \begin{cases} \mathrm{var}[\Delta(f \circ C) \mid C] = f'^2 \circ C \cdot \mathrm{var}[\Delta C \mid C] \\ \mathbb{E}[\Delta(f \circ C) \mid C] = f' \circ C \, \mathbb{E}[\Delta C \mid C] + \frac{1}{2} f'' \circ C \cdot \mathrm{var}[\Delta C \mid C]. \end{cases}$$

Let us introduce the two functions γ and a, defined by

$$\begin{aligned} \mathrm{var}[\Delta C \mid C] &= \gamma(C)\varepsilon^2 \\ \mathbb{E}[\Delta C \mid C] &= a(C)\varepsilon^2, \end{aligned}$$

where ε is a size parameter denoting the smallness of errors; (1) can then be written

$$(2) \qquad \begin{cases} \mathrm{var}[\Delta(f \circ C) \mid C] = f'^2 \circ C \cdot \gamma(C)\varepsilon^2 \\ \mathbb{E}[\Delta(f \circ C) \mid C] = f' \circ C \cdot a(C)\varepsilon^2 + \frac{1}{2} f'' \circ C \cdot \gamma(C)\varepsilon^2. \end{cases}$$

In examining the image probability space by C, i.e. the probability space

$$(\mathbb{R}, \mathcal{B}(\mathbb{R}), \mathbb{P}_C)$$

where $\mathbb{P}_C$ is the law of C. By virtue of the preceding we derive an operator Γ_C which, for any function f, provides the conditional variance of the error on $f \circ C$:

$$\varepsilon^2 \Gamma_C[f] \circ C = \operatorname{var}[\Delta(f \circ C) \mid C] = f'^2 \circ C \cdot \gamma(C)\varepsilon^2 \quad \mathbb{P}\text{-a.s.}$$

or, equivalently,

$$\varepsilon^2 \Gamma_C[f](x) = \operatorname{var}[\Delta(f \circ C) \mid C = x] = f'^2(x)\gamma(x)\varepsilon^2 \quad \text{for } \mathbb{P}_C\text{-a.e. } x.$$

The object $(\mathbb{R}, \mathcal{B}(\mathbb{R}), \mathbb{P}_C, \Gamma_C)$ with, in this case $\Gamma_C[f] = f'^2 \cdot \gamma$, suitably axiomatized will be called an *error structure* and Γ_C will be called the *quadratic error operator* of this error structure.

2.2. What happens when C is a two-dimensional random variable? Let us take an example.

Suppose a duration T_1 follows an exponential law of parameter 1 and is measured in such a manner that T_1 and its error can be modeled by the error structure

$$\begin{cases} S_1 = \big(\mathbb{R}_+, \mathcal{B}(\mathbb{R}_+), e^{-x} 1_{[0,\infty[}(x)\, dx, \Gamma_1\big) \\ \Gamma_1[f](x) = f'^2(x)\alpha^2 x^2 \end{cases}$$

which expresses the fact that

$$\operatorname{var}[\Delta T_1 \mid T_1] = \alpha^2 T_1^2 \varepsilon^2.$$

Similarly, suppose a duration T_2 following the same law is measured by another device such that T_2 and its error can be modeled by the following error structure:

$$\begin{cases} S_2 = \big(\mathbb{R}_+, \mathcal{B}(\mathbb{R}_+), e^{-y} 1_{[0,\infty[}(y)\, dy, \Gamma_2\big) \\ \Gamma_2[f](y) = f'^2(y)\beta^2 y^2. \end{cases}$$

In order to compute errors on functions of T_1 and T_2, hypotheses are required both on the joint law of T_1 and T_2 and on the correlation or uncorrelation of the errors.

a) Let us first suppose that pairs $(T_1, \Delta T_1)$ and $(T_2, \Delta T_2)$ are independent. Then the image probability space of (T_1, T_2) is

$$\big(\mathbb{R}_+^2, \mathcal{B}(\mathbb{R}_+^2), 1_{[0,\infty[}(x) 1_{[0,\infty[}(y) e^{-x-y}\, dx\, dy\big).$$

The error on a regular function F of T_1 and T_2 is

$$\begin{aligned}\Delta\big(F(T_1,T_2)\big) &= F_1'(T_1,T_2)\Delta T_1 + F_2'(T_1,T_2)\Delta T_2\\ &\quad+\frac{1}{2}F_{11}''(T_1,T_2)\Delta T_1^2 + F_{12}''(T_1,T_2)\Delta T_1\Delta T_2\\ &\quad+\frac{1}{2}F_{22}''(T_1,T_2)\Delta T_2^2 + \text{negligible terms}\end{aligned}$$

and, using assumptions A1 to A3, we obtain

$$\begin{aligned}\operatorname{var}[\Delta(F(T_1,T_2)) \mid T_1,T_2] &= \mathbb{E}[(\Delta(F(T_1,T_2)))^2 \mid T_1,T_2]\\ &= F_1'^2(T_1,T_2)\mathbb{E}[(\Delta T_1)^2 \mid T_1,T_2]\\ &\quad+2F_1'(T_1,T_2)F_2'(T_1,T_2)\mathbb{E}[\Delta T_1\Delta T_2 \mid T_1,T_2]\\ &\quad+F_2'^2(T_1,T_2)\mathbb{E}[(\Delta T_2)^2 \mid T_1,T_2].\end{aligned}$$

We use the following lemma (exercise):

Lemma I.1. *If the pairs (U_1,V_1) and (U_2,V_2) are independent, then*

$$\mathbb{E}[U_1,U_2 \mid V_1,V_2] = \mathbb{E}[U_1 \mid V_1]\cdot\mathbb{E}[U_2 \mid V_2].$$

Once again we obtain with A1 to A3:

$$\begin{aligned}\operatorname{var}[\Delta(F(T_1,T_2)) \mid T_1,T_2] &= F_1'^2(T_1,T_2)\operatorname{var}[\Delta T_1 \mid T_1]\\ &\quad+F_2'^2(T_1,T_2)\operatorname{var}[\Delta T_1 \mid T_2].\end{aligned}$$

In other words, the quadratic operator Γ of the error structure modeling T_1, T_2 and their errors

$$\big(\mathbb{R}_+^2,\ \mathcal{B}(\mathbb{R}_+^2),\ 1_{[0,\infty[}(x)1_{[0,\infty[}(y)e^{-x-y}\,dx\,dy,\ \Gamma\big)$$

satisfies

$$\Gamma[F](x,y) = \Gamma_1[F(\cdot,y)](x) + \Gamma_2[F(x,\cdot)](y).$$

If we consider that the conditional laws of errors are very concentrated Gaussian laws with dispersion matrix

$$M = \varepsilon^2\begin{pmatrix}\alpha^2x^2 & 0\\ 0 & \beta^2y^2\end{pmatrix},$$

hence with density

$$\frac{1}{2\pi}\frac{1}{\sqrt{\det M}}\exp\left\{-\frac{1}{2}(u\ \ v)M^{-1}\begin{pmatrix}u\\ v\end{pmatrix}\right\},$$

we may graphically represent errors by the elliptic level curves of these Gaussian densities of equations

$$(u \;\; v)M^{-1}\begin{pmatrix} u \\ v \end{pmatrix} = 1.$$

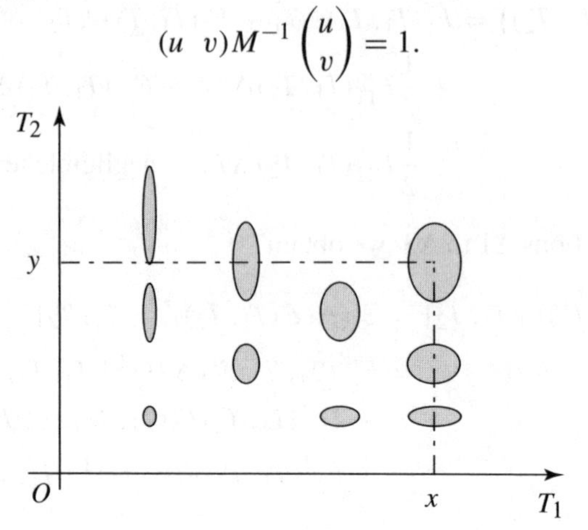

b) Let us now weaken the independence hypothesis by supposing T_1 and T_2 to be independent but their errors not. This assumption means that the quantity

$$\mathbb{E}[\Delta T_1 \Delta T_2 \mid T_1, T_2] - \mathbb{E}[\Delta T_1 \mid T_1, T_2]\mathbb{E}[\Delta T_2 \mid T_1, T_2],$$

which is always equal to

$$\mathbb{E}\big[(\Delta T_1 - \mathbb{E}[\Delta T_1 \mid T_1, T_2])(\Delta T_2 - \mathbb{E}[\Delta T_2 \mid T_1, T_2]) \mid T_1, T_2\big],$$

no longer vanishes, but remains a function of T_1 and T_2. This quantity is called the conditional covariance of ΔT_1 and ΔT_2 given T_1, T_2 and denoted by $\mathrm{cov}[(\Delta T_1, \Delta T_2) \mid T_1, T_2]$.

As an example, we can take

$$\mathrm{cov}[(\Delta T_1, \Delta T_2) \mid T_1, T_2] = \rho T_1 T_2 \varepsilon^2$$

with $\alpha^2\beta^2 - \rho^2 \geq 0$ so that the matrix

$$\begin{pmatrix} \mathrm{var}[\Delta T_1 \mid T_1, T_2] & \mathrm{cov}[(\Delta T_1, \Delta T_2) \mid T_1, T_2] \\ \mathrm{cov}[(\Delta T_1, \Delta T_2) \mid T_1, T_2] & \mathrm{var}[\Delta T_2 \mid T_1, T_2] \end{pmatrix} = \begin{pmatrix} \alpha^2 T_1^2 & \rho T_1 T_2 \\ \rho T_1 T_2 & \beta^2 T_2^2 \end{pmatrix} \varepsilon^2$$

is positive semi-definite, as is the case with any variance-covariance matrix.

If we were to compute as before the error on a regular function F of T_1, T_2, we would then obtain

$$\begin{aligned} &\mathrm{var}[\Delta(F(T_1, T_2)) \mid T_1, T_2] \\ &\quad = F_1'^2(T_1, T_2)\alpha^2 T_1^2 \varepsilon^2 + 2F_1'(T_1, T_2)F_2'(T_1, T_2)\rho T_1 T_2 \varepsilon^2 + F_2'^2(T_1, T_2)\beta^2 T_2^2 \varepsilon^2 \end{aligned}$$

and the quadratic operator is now

$$\Gamma[F](x,y) = F_1'^2(x,y)\alpha^2x^2 + 2F_1'(x,y)F_2'(x,y)\rho xy + F_2'^2(x,y)\beta^2y^2.$$

If, as in the preceding case, we consider that the conditional laws of errors are very concentrated Gaussian laws with dispersion matrix

$$M = \varepsilon^2\begin{pmatrix}\alpha^2x^2 & \rho xy\\ \rho xy & \beta^2y^2\end{pmatrix},$$

the elliptic level curves of these Gaussian densities with equation

$$(u \;\; v)M^{-1}\begin{pmatrix}u\\ v\end{pmatrix} = 1$$

may be parametrized by

$$\begin{pmatrix}u\\ v\end{pmatrix} = \sqrt{M}\begin{pmatrix}\cos\theta\\ \sin\theta\end{pmatrix},$$

where $\sqrt{M}$ is the symmetric positive square root of the matrix M. We see that

$$u^2+v^2 = (\cos\theta \;\; \sin\theta)M\begin{pmatrix}\cos\theta\\ \sin\theta\end{pmatrix} = \varepsilon^2\Gamma[T_1\cos\theta + T_2\sin\theta](x,y),$$

hence $\sqrt{u^2+v^2}$ is the standard deviation of the error in the direction θ.

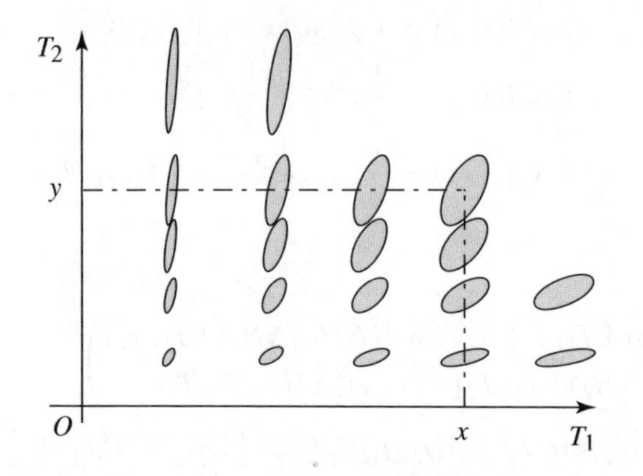

c) We can also abandon the hypothesis of independence of T_1 and T_2. The most general error structure on $(\mathbb{R}_+^2, \mathcal{B}(\mathbb{R}_+^2))$ would then be

$$\big(\mathbb{R}_+^2, \mathcal{B}(\mathbb{R}_+^2), \mu(dx,dy), \Gamma\big),$$

where μ is a probability measure and Γ is an operator of the form

$$\Gamma[F](x,y) = F_1'^2(x,y)a(x,y) + 2F_1'(x,y)F_2'(x,y)b(x,y) + F_2'^2(x,y)c(x,y)$$

where the matrix

$$\begin{pmatrix} a(x,y) & b(x,y) \\ b(x,y) & c(x,y) \end{pmatrix}$$

is positive semi-definite. Nevertheless, we will see further below that in order to achieve completely satisfactory error calculus, a link between the measure μ and the operator Γ will be necessary.

Exercise. Consider the error structure of Section 2.2.a):

$$\begin{cases} \left(\mathbb{R}_+^2, \mathcal{B}(\mathbb{R}_+^2), 1_{[0,\infty[}(x)1_{[0,\infty[}(y)e^{-x-y}\,dx\,dy, \Gamma\right) \\ \Gamma[F](x,y) = F_1'^2(x,y)\alpha^2x^2 + F_2'^2(x,y)\beta^2y^2 \end{cases}$$

and the random variable H with values in $\mathbb{R}^2$ defined by

$$H = (H_1, H_2) = \left(T_1 \wedge T_2, \frac{T_1 + T_2}{2}\right).$$

What is the conditional variance of the error on H?

Being bivariate, the random variable H possesses a bivariate error and we are thus seeking a 2×2-matrix.

Setting $F(x,y) = x \wedge y$, $G(x,y) = \frac{x+y}{2}$, we have

$$\begin{aligned} \Gamma[F](x,y) &= 1_{\{x \le y\}}\alpha^2x^2 + 1_{\{y \le x\}}\beta^2y^2 \\ \Gamma[G](x,y) &= \frac{1}{4}\alpha^2x^2 + \frac{1}{4}\beta^2y^2 \\ \Gamma[F,G](x,y) &= \frac{1}{2}1_{\{x \le y\}}\alpha^2x^2 + \frac{1}{2}1_{\{y \le x\}}\beta^2y^2 \end{aligned}$$

and eventually

$$\begin{aligned} &\begin{pmatrix} \operatorname{var}[\Delta H_1 \mid T_1, T_2] & \operatorname{cov}[(\Delta H_1, \Delta H_2) \mid T_1, T_2] \\ \operatorname{cov}[(\Delta H_1, \Delta H_2) \mid T_1, T_2] & \operatorname{var}[\Delta H_2 \mid T_1, T_2] \end{pmatrix} \\ &= \begin{pmatrix} 1_{\{T_1 \le T_2\}}\alpha^2T_1^2 + 1_{\{T_2 \le T_1\}}\beta^2T_2^2 & \frac{1}{2}1_{\{T_1 \le T_2\}}\alpha^2T_1^2 + \frac{1}{2}1_{\{T_2 \le T_1\}}\beta^2T_2^2 \\ \frac{1}{2}1_{\{T_1 \le T_2\}}\alpha^2T_1^2 + \frac{1}{2}1_{\{T_2 \le T_1\}}\beta^2T_2^2 & \frac{1}{4}\alpha^2T_1^2 + \frac{1}{4}\beta^2T_2^2 \end{pmatrix}. \end{aligned}$$

3 Intuitive notion of error structure

The preceding example shows that the quadratic error operator Γ naturally polarizes into a bilinear operator (as the covariance operator in probability theory), which is a first-order differential operator.

3.1. We thus adopt the following temporary definition of an *error structure*.

An error structure is a probability space equipped with an operator Γ acting upon random variables

$$(\Omega, \mathcal{X}, \mathbb{P}, \Gamma)$$

and satisfying the following properties:

a) *Symmetry*

$$\Gamma[F, G] = \Gamma[G, F];$$

b) *Bilinearity*

$$\Gamma\Big[\sum_i \lambda_i F_i, \sum_j \mu_j G_j\Big] = \sum_{ij} \lambda_i \mu_j \Gamma[F_i, G_j];$$

c) *Positivity*

$$\Gamma[F] = \Gamma[F, F] \geq 0;$$

d) *Functional calculus on regular functions*

$$\begin{aligned}\Gamma[\Phi(F_1, \dots, F_p), \Psi(G_1, \dots, G_q)]\\ = \sum_{i,j} \Phi_i'(F_1, \dots, F_p)\Psi_j'(G_1, \dots, G_q)\Gamma[F_i, G_j].\end{aligned}$$

3.2. In order to take in account the biases, we also have to introduce a bias operator A, a linear operator acting on regular functions through a second order functional calculus involving Γ:

$$\begin{aligned}A[\Phi(F_1, \dots, F_p)] = \sum_i \Phi_i'(F_1, \dots, F_p) A[F_i]\\ + \frac{1}{2}\sum_{ij} \Phi_{ij}''(F_1, \dots, F_p)\Gamma[F_i, F_j].\end{aligned}$$

Actually, the operator A will be yielded as a consequence of the probability space $(\Omega, \mathcal{X}, \mathbb{P})$ and the operator Γ. This fact needs the theory of operator semigroups which will be exposed in Chapter II.

3.3. Let us give an intuitive manner to pass from the classical probabilistic thought of errors to a modelisation by an error structure. We have to consider that

$$(\Omega, \mathcal{X}, \mathbb{P})$$

represents what can be obtained by experiment and that the errors are small and only known by their two first conditional moments with respect to the σ-field $\mathcal{X}$. Then, up to a size renormalization, we must think Γ and A as

$$\begin{aligned} \Gamma[X] &= \mathbb{E}[(\Delta X)^2 | \mathcal{X}] \\ A[X] &= \mathbb{E}[\Delta X | \mathcal{X}] \end{aligned}$$

where ΔX is the error on X. These two quantities have the same order of magnitude.

4 How to proceed with an error calculation

4.1. Suppose we are drawing a triangle with a graduated rule and a protractor: we take the polar angle of OA, say θ_1, and set $OA = \ell_1$; next we take the angle (OA, AB), say θ_2, and set $AB = \ell_2$.

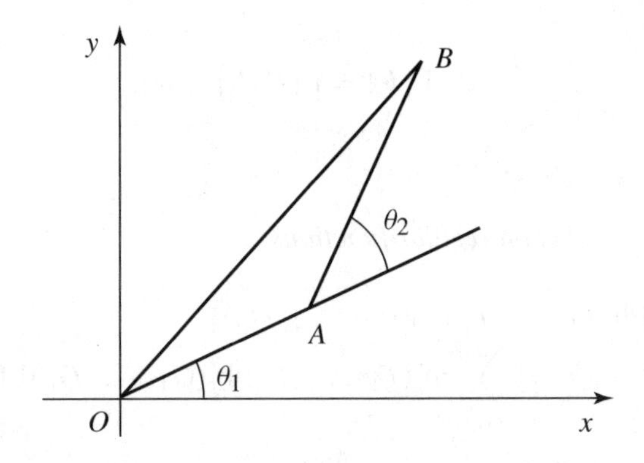

1) Select hypotheses on errors. ℓ_1, ℓ_2 and θ_1, θ_2 and their errors can be modeled as follows:

$$\left((0, L)^2 \times (0, \pi)^2, \mathcal{B}\big((0, L)^2 \times (0, \pi)^2\big), \frac{d\ell_1}{L}\frac{d\ell_2}{L}\frac{d\theta_1}{\pi}\frac{d\theta_2}{\pi}, \mathbb{D}, \Gamma\right)$$

where

$$\mathbb{D} = \left\{ f \in L^2\left(\frac{d\ell_1}{L}\frac{d\ell_2}{L}\frac{d\theta_1}{\pi}\frac{d\theta_2}{\pi}\right) : \frac{\partial f}{\partial \ell_1}, \frac{\partial f}{\partial \ell_2}, \frac{\partial f}{\partial \theta_1}, \frac{\partial f}{\partial \theta_2} \in L^2\left(\frac{d\ell_1}{L}\frac{d\ell_2}{L}\frac{d\theta_1}{\pi}\frac{d\theta_2}{\pi}\right)\right\}$$

and

$$\Gamma[f] = \ell_1^2 \left(\frac{\partial f}{\partial \ell_1}\right)^2 + \ell_1 \ell_2 \frac{\partial f}{\partial \ell_1}\frac{\partial f}{\partial \ell_2} + \ell_2^2 \left(\frac{\partial f}{\partial \ell_2}\right)^2 + \left(\frac{\partial f}{\partial \theta_1}\right)^2 + \frac{\partial f}{\partial \theta_1}\frac{\partial f}{\partial \theta_2} + \left(\frac{\partial f}{\partial \theta_2}\right)^2.$$

This quadratic error operator indicates that the errors on lengths ℓ_1, ℓ_2 are uncorrelated with those on angles θ_1, θ_2 (i.e. no term in $\frac{\partial f}{\partial \ell_i}\frac{\partial f}{\partial \theta_j}$). Such a hypothesis proves natural when measurements are conducted using different instruments. The bilinear operator associated with Γ is

$$\Gamma[f,g] = \ell_1^2 \frac{\partial f}{\partial \ell_1}\frac{\partial g}{\partial \ell_1} + \frac{1}{2}\ell_1\ell_2\left(\frac{\partial f}{\partial \ell_1}\frac{\partial g}{\partial \ell_2} + \frac{\partial f}{\partial \ell_2}\frac{\partial g}{\partial \ell_1}\right) + \ell_2^2\frac{\partial f}{\partial \ell_2}\frac{\partial g}{\partial \ell_2}$$

$$+ \frac{\partial f}{\partial \theta_1}\frac{\partial g}{\partial \theta_1} + \frac{1}{2}\left(\frac{\partial f}{\partial \theta_1}\frac{\partial g}{\partial \theta_2} + \frac{\partial f}{\partial \theta_2}\frac{\partial g}{\partial \theta_1}\right) + \frac{\partial f}{\partial \theta_2}\frac{\partial g}{\partial \theta_2}.$$

2) Compute the errors on significant quantities using functional calculus on Γ (Property 3d)). Take point B for instance:

$$\begin{aligned}
X_B &= \ell_1 \cos\theta_1 + \ell_2\cos(\theta_1+\theta_2), \quad Y_B = \ell_1\sin\theta_1 + \ell_2\sin(\theta_1+\theta_2)\\
\Gamma[X_B] &= \ell_1^2 + \ell_1\ell_2(\cos\theta_2 + 2\sin\theta_1\sin(\theta_1+\theta_2))\\
&\quad + \ell_2^2(1+2\sin^2(\theta_1+\theta_2))\\
\Gamma[Y_B] &= \ell_1^2 + \ell_1\ell_2(\cos\theta_2 + 2\cos\theta_1\cos(\theta_1+\theta_2))\\
&\quad + \ell_2^2(1+2\cos^2(\theta_1+\theta_2))\\
\Gamma[X_B, Y_B] &= -\ell_1\ell_2\sin(2\theta_1+\theta_2) - \ell_2^2\sin(2\theta_1+2\theta_2).
\end{aligned}$$

For the area of the triangle, the formula $\text{area}(OAB) = \frac{1}{2}\ell_1\ell_2\sin\theta_2$ yields

$$\Gamma[\text{area}(OAB)] = \frac{1}{4}\ell_1^2\ell_2^2(1+2\sin^2\theta_2).$$

The proportional error on the triangle area

$$\frac{(\Gamma[\text{area}(OAB)])^{1/2}}{\text{area}(OAB)} = \left(\frac{1}{\sin^2\theta_2}+2\right)^{1/2} \geq \sqrt{3}$$

reaches a minimum at $\theta_2 = \frac{\pi}{2}$ when the triangle is rectangular. From the equation $OB^2 = \ell_1^2 + 2\ell_1\ell_2\cos\theta_2 + \ell_2^2$ we obtain

$$\begin{aligned}
\Gamma[OB^2] &= 4\left[(\ell_1^2+\ell_2^2)^2 + 3(\ell_1^2+\ell_2^2)\ell_1\ell_2\cos\theta_2 + 2\ell_1^2\ell_2^2\cos^2\theta_2\right]\\
&= 4OB^2(OB^2 - \ell_1\ell_2\cos\theta_2)
\end{aligned}$$

and by $\Gamma[OB] = \frac{\Gamma[OB^2]}{4OB^2}$ we have

$$\frac{\Gamma[OB]}{OB^2} = 1 - \frac{\ell_1\ell_2\cos\theta_2}{OB^2},$$

thereby providing the result that the proportional error on OB is minimal when $\ell_1 = \ell_2$ and $\theta_2 = 0$. In this case $\frac{(\Gamma[OB])^{1/2}}{OB} = \frac{\sqrt{3}}{2}$.

5 Application: Partial integration for a Markov chain

Let (X_t) be a Markov process with values in $\mathbb{R}$ for the sake of simplicity. We are seeking to calculate by means of simulation the 1-potential of a bounded regular function f:

$$\mathbb{E}_x\left[\int_0^\infty e^{-t} f(X_t)\,dt\right]$$

and the derivative

$$\frac{d}{dx}\mathbb{E}_x\left[\int_0^\infty e^{-t} f(X_t)\,dt\right].$$

Suppose that the Markov chain (X_n^x) is a discrete approximation of (X_t) and simulated by

$$X_{n+1}^x = \Psi(X_n^x, U_{n+1}), \quad X_0^x = x, \tag{3}$$

where $U_1, U_2, \dots, U_n, \dots$ is a sequence of i.i.d. random variables uniformly distributed over the interval $[0, 1]$ representing the Monte Carlo samples. The 1-potential is then approximated by

$$P = \mathbb{E}\Big[\sum_{n=0}^{\infty} e^{-n\Delta t} f(X_n^x)\Delta t\Big]. \tag{4}$$

Let us now suppose that the first Monte Carlo sample U_1 is erroneous and represented by the following error structure:

$$\big([0, 1], \mathcal{B}([0, 1]), 1_{[0,1]}(x)\,dx, \Gamma\big)$$

with

$$\Gamma[h](x) = h'^2(x)x^2(1-x)^2.$$

Then, for regular functions h, k,

$$\int_0^1 \Gamma[h,k](x)\,dx = \int_0^1 h'(x)k'(x)x^2(1-x^2)\,dx$$

yields by partial integration

$$\int_0^1 \Gamma[h,k](x)\,dx = -\int_0^1 h(x)\big(k'(x)x^2(1-x)^2\big)'\,dx. \tag{5}$$

In other words, in our model $U_1, U_2, \dots, U_n, \dots$ only U_1 is erroneous and we have

$$\Gamma[U_1] = U_1^2(1-U_1)^2.$$

Hence by means of functional calculus (Property 3d)

$$\begin{aligned} &\Gamma[F(U_1,\dots,U_n,\dots), G(U_1,\dots,U_n,\dots)] \\ &\quad = F_1'(U_1,\dots,U_n,\dots)G_1'(U_1,\dots,U_n,\dots)U_1^2(1-U_1)^2 \end{aligned} \tag{6}$$

and (5) implies

$$\mathbb{E}\Gamma[F(U_1,\dots,U_n,\dots),G(U_1,\dots,U_n,\dots)]$$

(7)
$$= -\mathbb{E}\left[F(U_1,\dots,U_n,\dots)\frac{\partial}{\partial U_1}\left(\frac{\partial G}{\partial U_1}(U_1,\dots,U_n,\dots)U_1^2(1-U_1)^2\right)\right].$$

The derivative of interest to us then becomes

$$\frac{dP}{dx} = \mathbb{E}\left[\sum_{n=0}^{\infty} e^{-n\Delta t}\frac{\partial(f(X_n^x))}{\partial x}\Delta t\right]$$

and by the representation in (3)

(8)
$$\frac{\partial f(X_n^x)}{\partial x} = f'(X_n^x)\prod_{i=0}^{n-1}\Psi_1'(X_i^x, U_{i+1}).$$

However, we can observe that

(9)
$$\frac{\partial f(X_n^x)}{\partial U_1} = f'(X_n^x)\prod_{i=1}^{n-1}\Psi_1'(X_i^x, U_{i+1})\Psi_2'(x, U_1),$$

and comparing (8) with (9) yields

$$\frac{dP}{dx} = \mathbb{E}\left[\left(\sum_{n=0}^{\infty} e^{-n\Delta t}\Delta t\frac{\partial f(X_n^x)}{\partial U_1}\right)\frac{\Psi_1'(x,U_1)}{\Psi_2'(x,U_1)}\right].$$

This expression can be treated by applying formula (6) with

$$F_1'(U_1,\dots,U_n,\dots) = \sum_{n=0}^{\infty} e^{-n\Delta t}\Delta t\frac{\partial f(X_n^x)}{\partial U_1}$$

$$G_1'(U_1,\dots,U_n,\dots)U_1^2(1-U_1)^2 = \frac{\Psi_1'(x,U_1)}{\Psi_2'(x,U_1)}.$$

This gives

(10)
$$\frac{dP}{dx} = -\mathbb{E}\left[\left(\sum_{n=0}^{\infty} e^{-n\Delta t}\Delta t f(X_n^x)\right)\frac{\partial}{\partial U_1}\left(\frac{\Psi_1'(x,U_1)}{\Psi_2'(x,U_1)}\right)\right].$$

Formula (10) is a typical integration by parts formula, useful in Monte Carlo simulation when simultaneously dealing with several functions f.

One aim of error calculus theory is to generalize such integration by parts formulae to more complex contexts.

We must now focus on making such error calculations more rigorous. This process will be carried out in the following chapters using a powerful mathematical toolbox, the theory of Dirichlet forms. The benefit consists of the possibility of performing error calculations in infinite dimensional models, as is typical in stochastic analysis and in mathematical finance in particular. Other advantages will be provided thanks to the strength of rigorous arguments.

The notion of error structure will be axiomatized in Chapter III. A comparison of error calculus based on error structures, i.e. using Dirichlet forms, with other methods will be performed in Chapter V, Section 1.2. Error calculus will be described as an extension of probability theory. In particular, if we are focusing on the sensitivity of a model to a parameter, use of this theory necessitates for this parameter to be randomized first. and can then be considered erroneous. As we will see further below, the choice of this *a priori* law is not as crucial as may be thought provided our interest lies solely in the error variances. The *a priori* law is important when examining error biases.

Let us contribute some historical remarks on *a priori* laws.

Appendix. Historical comment: the benefit of randomizing physical or natural quantities

The founders of the so-called *classical error theory* at the beginning of the 19th century, i.e. Legendre, Laplace, and Gauss, were the first to develop a rigorous argument in this area. One example is Gauss' famous proof of the 'law of errors'. Gauss showed that if having taken measurements x_i, the arithmetic average $\frac{1}{n}\sum_{i=1}^n x_i$ is the value we prefer as the best one, then (with additional assumptions, some of which are implicit and have been pointed out later by other authors) the errors necessarily obey a normal law, and the arithmetic average is both the most likely value and the one generated from the least squares method.

Gauss tackled this question in the following way. He first assumed – we will return to this idea later on – that the quantity to be measured is random and can vary within the domain of the measurement device according to an *a priori* law. In more modern language, let X be this random variable and μ its law. The results of the measurement operations are other random variables $X_1, \dots, X_n$ and Gauss assumes that:

a) the conditional law of X_i given X is of the form

$$\mathbb{P}\{X_i \in E \mid X = x\} = \int_E \varphi(x_1 - x)\, dx_1,$$

b) the variables $X_1, \dots, X_n$ are conditionally independent given X.

He then easily computed the conditional law of X given the measurement results: it displays a density with respect to μ. This density being maximized at the arithmetic average, he obtains:

$$\frac{\varphi'(t-x)}{\varphi(t-x)} = a(t-x),$$

hence:

$$\varphi(t-x) = \frac{1}{\sqrt{2\pi\sigma^2}} \exp\left(-\frac{(t-x)^2}{2\sigma^2}\right).$$

In Poincaré's *Calcul des Probabilités* at the end of the century, it is likely that Gauss' argument is the most clearly explained, in that Poincaré attempted to both present all hypotheses explicitly and generalize the proof [1]. He studied the case where the conditional law of X_1 given X is no longer $\varphi(y-x)\,dy$ but of the more general form $\varphi(y,x)\,dy$. This led Poincaré to suggest that the measurements could be independent while the errors need not be, when performed with the same instrument. He did not develop any new mathematical formalism for this idea, but emphasized the advantage of assuming small errors: This allows Gauss' argument for the normal law to become compatible with nonlinear changes of variables and to be carried out by differential calculus. This focus is central to the field of *error calculus*.

Twelve years after his demonstration that led to the normal law, Gauss became interested in the propagation of errors and hence must be considered as the founder of error calculus. In *Theoria Combinationis* (1821) he states the following problem. Given a quantity $U = F(V_1, V_2, \dots)$ function of the erroneous quantities $V_1, V_2, \dots$, compute the potential quadratic error to expect on U, with the quadratic errors σ_1^2, $\sigma_2^2, \dots$ on $V_1, V_2, \dots$ being known and assumed to be small and independent. His response consisted of the following formula:

$$\text{(11)} \qquad \sigma_U^2 = \left(\frac{\partial F}{\partial V_1}\right)^2 \sigma_1^2 + \left(\frac{\partial F}{\partial V_2}\right)^2 \sigma_2^2 + \cdots.$$

He also provided the covariance between the error on U and the error of another function of the V_i's.

Formula (11) displays a property that enhances its attractiveness in several respects over other formulae encountered in textbooks throughout the 19th and 20th centuries: it has a *coherence* property. With a formula such as

$$\text{(12)} \qquad \sigma_U = \left|\frac{\partial F}{\partial V_1}\right| \sigma_1 + \left|\frac{\partial F}{\partial V_2}\right| \sigma_2 + \cdots$$

errors may depend on the manner in which the function F is written; in dimension 2 we can already observe that if we write the identity map as the composition of an injective linear map with its inverse, we are increasing the errors (a situation which is hardly acceptable).

[1] It is regarding this 'law of errors' that Poincaré wrote: "Everybody believes in it because experimenters imagine it to be a theorem of mathematics while mathematicians take it as experimental fact."

This difficulty however does not occur in Gauss' calculus. Introducing the operator

$$L = \frac{1}{2}\sigma_1^2 \frac{\partial^2}{\partial V_1^2} + \frac{1}{2}\sigma_2^2 \frac{\partial^2}{\partial V_2^2} + \cdots$$

and supposing the functions to be smooth, we remark that formula (11) can be written as follows:

$$\sigma_U^2 = LF^2 - 2FLF.$$

The coherence of this calculus follows from the coherence of the transport of a differential operator by a function: if L is such an operator and u and v injective regular maps, by denoting the operator $\varphi \to L(\varphi \circ u) \circ u^{-1}$ by $\theta_u L$, we then have $\theta_{v \circ u} L = \theta_v(\theta_u L)$.

The errors on $V_1, V_2, \dots$ are not necessarily supposed to be independent or constant and may depend on $V_1, V_2, \dots$ Considering a field of positive symmetric matrices $(\sigma_{ij}(v_1, v_2, \dots))$ on $\mathbb{R}^d$ to represent the conditional variances and covariances of the errors on $V_1, V_2, \dots$ given values $v_1, v_2, \dots$ of $V_1, V_2, \dots$, then the error of $U = F(V_1, V_2, \dots)$ given values $v_1, v_2, \dots$ of $V_1, V_2, \dots$ is

$$\sigma_F^2 = \sum_{ij} \frac{\partial F}{\partial v_i}(v_1, v_2, \dots) \frac{\partial F}{\partial v_j}(v_1, v_2, \dots) \sigma_{ij}(v_1, v_2, \dots) \tag{13}$$

which depends solely on F as a mapping.

Randomization has also been shown to be very useful in decision theory. The Bayesian methods within the statistical decision of A. Wald allow for optimization procedures thanks to the existence of an *a priori* law of probability.

In game theory, major advances have been made by Von Neumann through considering randomized policies.

For physical systems, E. Hopf (1934) has shown that for a large class of dynamic systems, time evolution gives rise to a special invariant measure on the state space and he gave explicit convergence theorems to this measure. We shall return to this theory in Chapter VIII.

Bibliography for Chapter I

N. Bouleau, *Probabilités de l'Ingénieur*, Hermann, Paris, 2002.

N. Bouleau, Calcul d'erreur complet lipschitzien et formes de Dirichlet, *J. Math. Pures Appl.* **80** (2001), 961–976.

L. Breiman, *Probability*, Addison-Wesley, 1968.

W. Feller, *An Introduction to Probability Theory and Its Applications*, Vol. 1, Wiley, 1950.

E. Hopf, On causality, statistics and probability, *J. Math. Phys.* **13** (1934), 51–102.

V. Ventsel and L. Ovtcharov, *Problèmes appliqués de la théorie des probabilités*, Ed., Mir, (1988), (théorie de la linéarisation, p. 200 *et seq.* and 247 *et seq.*).

Chapter II

Strongly-continuous semigroups and Dirichlet forms

In this chapter, we will account for the basic mathematical objects on which the theory of error structures has been built. We will be aiming for simplicity herein. After the main arguments concerning semigroups on Banach spaces have been stated, and this requires very little preliminary knowledge, the notion and properties of Dirichlet forms will be introduced for a special case that still follows a general reasoning.

1 Strongly-continuous contraction semigroups on a Banach space

Let B be a Banach space with norm $\|\cdot\|$.

Definition II.1. A family $\left(P_t\right)_{t\geq 0}$ of linear operators on B satisfying

1) $P_0 = I$ (identity), $P_{t+s} = P_t P_s$, P_t *is contracting* ($\|P_t x\| \leq \|x\| \quad \forall x \in B$),

2) $\lim_{t\to 0} P_t x = x \quad \forall x \in B$

will be called a *strongly-continuous contraction semigroup* on B.

Hereafter in this chapter, $\left(P_t\right)_{t\geq 0}$ will be a strongly-continuous contraction semigroup on B.

Exercise. Show that for every $x \in B$, the application $t \to P_t x$ is continuous from $\mathbb{R}_+$ into B.

Examples. a) Let P be a Fellerian probability kernel on $\mathbb{R}^d$, i.e. $P(x, dy)$ is a transition probability such that $\forall f \in \mathcal{C}_0\left(\mathbb{R}^d\right)$ (i.e. the space of continuous real functions on $\mathbb{R}^d$ vanishing at infinity) $Pf = \int f(y) P(x, dy)$ belongs to $\mathcal{C}_0\left(\mathbb{R}^d\right)$, then $P_t = \exp\{\lambda t(P - I)\}$ is a strongly-continuous contraction semigroup on $\mathcal{C}_0\left(\mathbb{R}^d\right)$ with the uniform norm $\|\cdot\|_\infty$. In addition, $P_t f \geq 0$ for $f \geq 0$. In this case, $\lim_{t\to 0} P_t = I$ in the sense of operators norm, i.e.:

$$\lim_{t\to 0} \sup_{\|x\|\leq 1} \left\| P_t x - x \right\| = 0.$$

This property is specific and related to the fact that $P - I$ is a bounded operator.

b) On $B = \mathcal{C}_0(\mathbb{R}^d)$ with the uniform norm let us define $P_t f(x) = f(x + kt)$, $k \in \mathbb{R}^d$, then $(P_t)_{t\geq 0}$ is a strongly-continuous contraction semigroup. The same holds if P_t acts on $L^p(\mathbb{R}^d)$ $1 \leq p < +\infty$ [to prove this assertion, use the fact that continuous functions with compact support are dense in $L^p(\mathbb{R}^d)$ $1 \leq p < +\infty$].

Definition II.2. The generator of $(P_t)_{t\geq 0}$ is the operator A with domain $\mathcal{D}A$ defined by

$$\mathcal{D}A = \left\{ x \in B\colon \lim_{t\downarrow 0} \frac{P_t x - x}{t} \textit{ exists in } B \right\},$$

and for $x \in \mathcal{D}A$

$$Ax = \lim_{t\to 0} \frac{P_t x - x}{t}.$$

We will need some elementary properties of integration of Banach-valued functions. We will only consider *continuous* functions from $\mathbb{R}$ into B so that integrals can be constructed in the Riemann sense. If $F\colon \mathbb{R} \to B$ is continuous then the following holds:

(i) for any continuous linear operator L on B

$$L\left(\int_a^b F(t)\,dt\right) = \int_a^b L(F(t))\,dt;$$

(ii)

$$\left\| \int_a^b F(t)\,dt \right\| \leq \int_a^b \|F(t)\|\,dt;$$

(iii) if F is C^1 then

$$\int_a^b \frac{dF}{dt}(s)\,ds = F(b) - F(a).$$

See Rudin (1973, Chapter 3) for complementary details.

Proposition II.3. *Let $x \in \mathcal{D}A$. Then*

1) $\forall t > 0 \quad P_t x \in \mathcal{D}A$ *and* $AP_t x = P_t Ax$,

2) *the map* $t \to P_t x$ *is differentiable with continuous derivative and*

$$\frac{d}{ds} P_s x \Big|_{s=t} = AP_t x = P_t Ax,$$

3) $P_t x - x = \displaystyle\int_0^t P_s Ax\,ds.$

Proof. 1) Observing $\lim_{s\to 0}\frac{1}{s}\big(P_s(P_t x) - P_t x\big)$, which is equal to

$$\lim_{s\to 0} P_t\left(\frac{1}{s}\big(P_s(x) - x\big)\right) = P_t A x,$$

shows that $P_t x \in \mathcal{D}A$ and $AP_t x = P_t A x$.

2) Thus $t \to P_t x$ admits $P_t A x$ as right derivative.

It is now also the left derivative, since

$$\lim_{s\downarrow 0}\frac{P_{t-s}x - P_t x}{-s} = P_{t-s}\frac{x - P_s x}{-s} = P_{t-s}\left[\frac{x - P_s x}{-s} - Ax\right] + P_{t-s}Ax,$$

and the inequality

$$\left\| P_{t-s}\left[\frac{x - P_s x}{-s} - Ax\right]\right\| \le \left\|\frac{x - P_s x}{-s} - Ax\right\| \xrightarrow[s\downarrow 0]{} 0$$

and the strong continuity of (P_t) yield the result.

Point 3) follows from 2) by means of property (iii). ◇

Proposition II.4. *$\mathcal{D}A$ is dense in B, and A is a closed operator (i.e. if a sequence $x_n \in \mathcal{D}A$ is such that $x_n \to x$ and $Ax_n \to y$ as $n \uparrow \infty$, then $x \in \mathcal{D}A$ and $Ax = y$).*

Proof. a) Let us introduce the bounded operators A_h and B_s:

$$A_h = \frac{P_h - I}{h} \qquad h > 0,$$

$$B_s = \frac{1}{s}\int_0^s P_t\,dt \quad s > 0.$$

A_h and B_s obviously commute, and furthermore

$$A_h B_s x = B_s A_h x = \frac{1}{sh}\left[\int_0^{s+h} - \int_0^h - \int_0^s P_u x\,du\right] = B_h A_s x = A_s B_h x.$$

In order to check whether $B_s x$ belongs to $\mathcal{D}A$, we have $A_h B_s x = B_h A_s x \xrightarrow[h\to 0]{} A_s x$, which proves $B_s x \in \mathcal{D}A$ and $AB_s x = A_s x$. Now $B_s x \to x$ as $s \to 0$, hence $\mathcal{D}A$ is dense.

b) Let x_n be a sequence in $\mathcal{D}A$ such that $x_n \to x$ and $Ax_n \to y$.

Noting that $\forall z \in \mathcal{D}A$

$$B_s A z = \lim_{h\downarrow 0} B_s A_h z = \lim_{h\downarrow 0} B_h A_s z = A_s z,$$

we obtain

$$B_s y = \lim_n B_s A x_n = \lim_n A_s x_n = A_s x.$$

By making $s \to 0$, the extreme terms of this equality show that $x \in \mathcal{D}A$ and $Ax = y$. $\diamond$

Exercise II.5. Let $f \in \mathcal{D}(]0, \infty[)$, i.e. f is infinitely derivable with compact support in $]0, \infty[$, and let us define

$$P_f x = \int_0^\infty f(t) P_t x \, dt.$$

a) Studying the limit as $h \to 0$ of $A_h P_f x$ reveals that $P_f x \in \mathcal{D}A$ and

$$A P_f x = -P_{f'} x.$$

b) Let us define

$$\mathcal{D}A^2 = \{x \in \mathcal{D}A \colon Ax \in \mathcal{D}A\} \quad \text{and} \quad \mathcal{D}A^n = \{x \in \mathcal{D}A^{n-1} \colon Ax \in \mathcal{D}A^{n-1}\}.$$

Show that

$$P_f x \in \bigcap_n \mathcal{D}A^n.$$

c) Considering a sequence $f_n \in \mathcal{D}(]0, \infty[)$ such that $f_n \geq 0$, $\int_0^\infty f_n(t)\, dt = 1$, support $(f_n) \to \{0\}$, show that $P_{f_n} x \to x$, and therefore $\bigcap_n \mathcal{D}A^n$ is dense.

Taking example b) after definition II.1 leads to the well-known fact that infinitely-derivable functions with derivatives in C_0 are dense in C_0.

For complements on this section, see Yosida (1974, Chapter IX) and Rudin (1973, Chapter 13).

2 The Ornstein–Uhlenbeck semigroup on $\mathbb{R}$ and the associated Dirichlet form

Let $m = \frac{1}{\sqrt{2\pi}} e^{-\frac{y^2}{2}}\, dy$ be the reduced normal law on $\mathbb{R}$. The following properties are straightforward to prove step-by-step.

For $f \in L^2(\mathbb{R}, m)$ and $t \geq 0$

$$\int f^2\big(\sqrt{e^{-t}}\, x + \sqrt{1 - e^{-t}}\, y\big) dm(x)\, dm(y) = \int f^2 \, dm. \tag{1}$$

Indeed the first member is also $\mathbb{E}\big[f^2\big(\sqrt{e^{-t}}\, X + \sqrt{1 - e^{-t}}\, Y\big)\big]$ where X and Y are independent reduced Gaussian variables, the law of $\sqrt{e^{-t}}\, X + \sqrt{1 - e^{-t}}\, Y$ is m, which provides the required equality. $\diamond$

As a consequence, the map $(x, y) \to f\big(\sqrt{e^{-t}}\, x + \sqrt{1 - e^{-t}}\, y\big)$ belongs to $L^2(m \times m) \subset L^1(m \times m)$, hence the operator P_t defined via

$$P_t f(x) = \int f\big(\sqrt{e^{-t}}\, x + \sqrt{1 - e^{-t}}\, y\big)\, dm(y)$$

maps $L^2(\mathbb{R}, m)$ into $L^1(\mathbb{R}, m)$. In fact:

(2) *P_t is a linear operator from $L^2(\mathbb{R}, m)$ into itself with norm* 1.

Indeed, we remark

$$\int (P_t f)^2 \, dm \leq \int P_t(f^2) \, dm = \int f^2 \, dm$$

and $P_t 1 = 1$. ◇

With a similar argument as that in (1), we obtain

$$P_t P_s = P_{s+t}. \tag{3}$$

Proof. Let f be in $L^2(\mathbb{R}, m)$. The property easily stems from the equality

$$P_s f(x) = \mathbb{E}\big[f\big(\sqrt{e^{-s}}\, x + \sqrt{1 - e^{-s}}\, Y\big)\big]$$

where $Y \sim \mathcal{N}(0, 1)$, in using the fact that the sum of two independent Gaussian variables is a Gaussian variable whose variance the sum of the variances. ◇

We can then prove:

(4) *$(P_t)_{t \geq 0}$ is a strongly-continuous contraction semigroup on $L^2(\mathbb{R}, m)$.*
It is called the Ornstein–Uhlenbeck semigroup in dimension one.

From the definition of P_t, it follows by dominated convergence that for f bounded and continuous $(f \in C_b)$,

$$P_t f(x) \underset{t \downarrow 0}{\longrightarrow} f(x) \quad \forall x.$$

Hence dominated convergence also yields

$$\int (P_t f - f)^2 dm \underset{t \downarrow 0}{\longrightarrow} 0.$$

This result now extends from C_b to $L^2(\mathbb{R}, m)$ by density using $\|P_t\| \leq 1$. ◇

Let $f, g \in L^2(\mathbb{R}, m)$,

$$\int f \cdot P_t g \, dm = \int (P_t f) \cdot g \, dm. \tag{5}$$

Proof. The first member can be written as follows:

$$\mathbb{E}\big[f(X) g\big(\sqrt{e^{-t}}\, X + \sqrt{1 - e^{-t}}\, Y\big)\big]$$

where X and Y are independent reduced Gaussian.

Setting $Z = \sqrt{e^{-t}}\, X + \sqrt{1-e^{-t}}\, Y$, we have

$$\int f \cdot P_t g \, dm = \mathbb{E}[f(X)g(Z)],$$

where (X, Z) is a pair of reduced Gaussian variables with covariance $\sqrt{e^{-t}}$. This property is symmetric and we may now substitute (Z, X) for (X, Z) which yields (5). ◇

Let us define the bilinear form on $L^2(\mathbb{R}, m) \times L^2(\mathbb{R}, m)$,

$$\mathcal{E}_t[f, g] = \frac{1}{t}\langle f - P_t f, g\rangle,$$

and the associated quadratic form on $L^2(\mathbb{R}, m)$,

$$\mathcal{E}_t[f] = \frac{1}{t}\langle f - P_t, f, f\rangle$$

where $\langle\cdot,\cdot\rangle$ is the scalar product on $L^2(\mathbb{R}, m)$.

Then

(6) $$\mathcal{E}_t[f] \geq 0.$$

This expression stems from

$$\mathcal{E}_t[f] = \frac{1}{t}\left(\|f\|^2 - \|P_{\frac{t}{2}} f\|^2\right)$$

and

$$\|P_{\frac{t}{2}} f\| \leq \|f\|.$$

(Here $\|\cdot\|$ is the $L^2(\mathbb{R}, m)$-norm.) ◇

As a consequence of (4) we can denote $(A, \mathcal{D}A)$ the generator of (P_t).

(7) *For $g \in \mathcal{D}A$, the mapping $t \to \langle P_t g, g\rangle$ is convex.*

We know that for $g \in \mathcal{D}A$, $t \to P_t g$ is differentiable, and using (5)

$$\frac{d}{dt}\langle P_t g, g\rangle = \langle P_t A g, g\rangle = \langle P_t g, A g\rangle.$$

We can then derive once again

$$\frac{d^2}{dt^2}\langle P_t g, g\rangle = \langle P_t A g, A g\rangle = \langle P_{\frac{t}{2}} A g, P_{\frac{t}{2}} A g\rangle$$

which is positive. ◇

It follows that for $f \in L^2(m)$

(8) *the mapping* $t \to \mathcal{E}_t[f]$ *is decreasing* ($\mathcal{E}_t[f]$ *increases as t decreases*).

$\mathcal{D}A$ is dense in $L^2(\mathbb{R}, m)$ and the simple limit of a convex function is convex, hence $\varphi\colon t \to \langle P_t f, f\rangle$ is convex for every $f \in L^2(\mathbb{R}, m)$. Thus $\frac{1}{t}(\varphi(t) - \varphi(0))$ increases when t increases, that gives (8). ◇

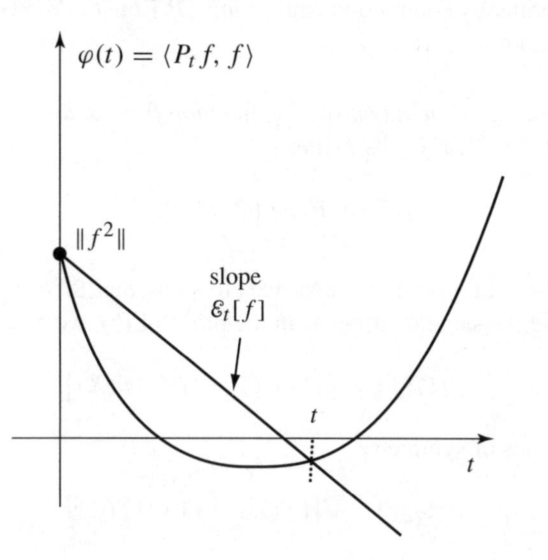

Let us define

$$\mathbb{D} = \left\{ f \in L^2(\mathbb{R}, m)\colon \lim_{t \to 0} \frac{1}{t}\langle f - P_t f, f\rangle \text{ exists} \right\}$$

$$\mathcal{E}[f] = \lim_{t \to 0} \frac{1}{t}\langle f - P_t f, f\rangle \ \text{ for } f \in \mathbb{D}.$$

The bilinearity of $\mathcal{E}_t$,

$$\mathcal{E}_t[\lambda f + \mu g] = \lambda^2 \mathcal{E}_t[f] + 2\lambda\mu\mathcal{E}_t[f, g] + \mu^2\mathcal{E}_t[g],$$

and the positivity (6) imply the classical inequalities

$$\mathcal{E}_t[f, g]^2 \le \mathcal{E}_t[f]\mathcal{E}_t[g]$$
$$\mathcal{E}_t[f + g]^{1/2} \le \mathcal{E}_t[f]^{1/2} + \mathcal{E}_t[g]^{1/2}.$$

Setting $t \to 0$ yields the same inequalities for $\mathcal{E}$, and $\mathbb{D}$ *is a vector space*.

Let us remark that $\mathbb{D}$ is dense in $L^2(\mathbb{R}, m)$ since $\mathcal{D}A \subset \mathbb{D}$ and for $f \in \mathcal{D}A$,

$$\mathcal{E}[f] = \langle -Af, f\rangle.$$

We polarize $\mathcal{E}$ as usual: for $f, g \in \mathbb{D}$, we define

$$\begin{aligned}\mathcal{E}[f, g] &= \frac{1}{2}(\mathcal{E}[f+g] - \mathcal{E}[f] - \mathcal{E}[g]) \\ &= \lim_{t \downarrow 0}\left\langle \frac{f - P_t f}{t}, g \right\rangle = \lim_{t \downarrow 0}\left\langle f, \frac{g - P_t g}{t} \right\rangle.\end{aligned}$$

The form $\mathcal{E}$ with domain $\mathbb{D}$ is called the *Dirichlet form* associated with the symmetric strongly-continuous contraction semigroup (P_t) on $L^2(\mathbb{R}, m)$ and possesses the following important property:

Proposition II.6. *Let F be a contracting function from $\mathbb{R}$ into $\mathbb{R}$ ($|F(x) - F(y)| \leq |x - y|$), then if $f \in \mathbb{D}$, $F \circ f \in \mathbb{D}$ and*

$$\mathcal{E}[F \circ f] \leq \mathcal{E}[f].$$

Proof. From the definition of $\mathcal{E}$, it suffices to show that $\mathcal{E}_t[F \circ f] \leq \mathcal{E}_t[f]$ $\forall f \in L^2(\mathbb{R}, m)$. Using the same notation as in the proof of (5), we now obtain

$$t\mathcal{E}_t[f] = \mathbb{E}\big[\big(f(X) - f(Z)\big)f(X)\big]$$

moreover by means of symmetry

$$t\mathcal{E}_t[f] = \mathbb{E}\big[\big(f(Z) - f(X)\big)f(Z)\big]$$

and taking the half-sum, we obtain

$$t\mathcal{E}_t[f] = \frac{1}{2}\mathbb{E}\big[\big(f(X) - f(Z)\big)^2\big].$$

The property $\mathcal{E}_t[F \circ f] \leq \mathcal{E}_t[f]$ is evident on this formula which concludes the proof. ◇

Similarly,

Proposition II.7. *Let $F : \mathbb{R}^n \to \mathbb{R}$ be a contraction in the following sense:*

$$|F(x) - F(y)| \leq \sum_{i=1}^{n} |x_i - y_i|.$$

Then $\forall f = (f_1, \ldots, f_n) \in \mathbb{D}^n$, $F \circ f \in \mathbb{D}$ and

$$\sqrt{\mathcal{E}(F \circ f)} \leq \sum_i \sqrt{\mathcal{E}[f_i]}.$$

Proof. It suffices to show that for every $f \in L^2(\mathbb{R}, m)^n$

$$\sqrt{\mathcal{E}_t[F \circ f]} \le \sum_i \sqrt{\mathcal{E}_t[f_i]}.$$

Now

$$\mathcal{E}_t[F \circ f] = \frac{1}{2t}\mathbb{E}\big[\big(F \circ f(X) - F \circ f(Y)\big)^2\big],$$

$$\begin{aligned}\sqrt{2t\mathcal{E}_t[F \circ f]} &= \|F \circ f(X) - F \circ f(Y)\|_{L^2} \le \Big\| \sum_i |f_i(X) - f_i(Y)| \Big\|_2 \\ &\le \sum_i \big\| f_i(X) - f_i(Y) \big\|_{L^2} = \sum_i \sqrt{2t\mathcal{E}_t[f_i]}.\end{aligned}$$

◇

Proposition II.8. *$\mathcal{E}$ is closed with dense domain.*

Proof. We already know that the domain $\mathbb{D}$ is dense.

The closedness of $\mathcal{E}$ means that $\mathbb{D}$, equipped with the norm

$$\|f\|_{\mathbb{D}} = \big(\mathcal{E}[f] + \|f\|^2_{L^2(m)}\big)^{1/2},$$

is complete, i.e. is a Hilbert space. For this it is to show that as soon as (f_n) is a Cauchy sequence in $(\mathbb{D}, \|\cdot\|_{\mathbb{D}})$, there exists an $f \in \mathbb{D}$ such that $\|f_n - f\|_{\mathbb{D}} \to 0$.

Let (f_n) be a Cauchy sequence in $\mathbb{D}$. By the definition of the norm $\|\cdot\|_{\mathbb{D}}$, (f_n) is also $L^2(m)$-Cauchy and an $f \in L^2(m)$ exists such that $f_n \to f$ in $L^2(m)$. Then

$$\begin{aligned}\mathcal{E}[f] &= \lim_{t \downarrow 0} \uparrow \mathcal{E}_t[f] \\ \mathcal{E}_t[f] &= \lim_n \mathcal{E}_t[f_n] \le \sup_n \mathcal{E}\big[f_n\big] < +\infty\end{aligned}$$

as a Cauchy sequence is necessarily bounded. Hence $\mathcal{E}[f] < +\infty$, i.e. $f \in \mathbb{D}$.

$$\begin{aligned}\mathcal{E}\big[f_n - f\big] &= \lim_t \mathcal{E}_t\big[f_n - f\big] \\ &= \lim_t \lim_m \mathcal{E}_t\big[f_n - f_m\big] = \lim_t \overline{\lim_m}\ \mathcal{E}_t\big[f_n - f_m\big] \\ &\le \overline{\lim_m}\ \mathcal{E}\big[f_n - f_m\big].\end{aligned}$$

This expression can now be made as small as desired for large n, since (f_n) is $\|\cdot\|_{\mathbb{D}}$-Cauchy. ◇

Proposition II.9. *$\mathbb{D} \cap L^\infty(m)$ is an algebra and*

$$\big(\mathcal{E}[fg]\big)^{1/2} \le \|f\|_\infty\big(\mathcal{E}[g]\big)^{1/2} + \|g\|_\infty\big(\mathcal{E}[f]\big)^{1/2}$$

$\forall f, g \in \mathbb{D} \cap L^\infty(m)$

Proof. By homogeneity, we can suppose that $\|f\|_\infty = \|g\|_\infty = 1$. Let $\varphi(x) = (x \wedge 1) \vee (-1)$ and $F(x, y) = \varphi(x)\varphi(y)$. F is a contraction, $fg = F(f, g)$ and the result stems from Proposition II.7. ◇

Probabilistic interpretation

Consider the following stochastic differential equation:

$$dX_t = dB_t - \frac{1}{2} X_t \, dt, \tag{9}$$

where B_t is a standard real Brownian motion. Noting that (9) can be written $d\big(e^{\frac{t}{2}} X_t\big) = e^{\frac{t}{2}} B_t$ gives

$$e^{\frac{t}{2}} X_t = X_0 + \int_0^t e^{\frac{s}{2}} \, dB_s.$$

The transition semigroup (P_t) of the associated Markov process is given by the expectation of $f(X_t)$ starting at x:

$$P_t f(x) = \mathbb{E}_x\big[f(X_t)\big].$$

$e^{\frac{t}{2}} X_t$ follows a normal law centered at x with variance $\int_0^t e^s \, ds = e^t - 1$, hence X_t follows a normal law centered at $xe^{-\frac{t}{2}}$ with variance $(e^t - 1)e^{-t} = 1 - e^{-t}$. In other words, (P_t) is the Ornstein–Uhlenbeck semigroup.

Let f be a C^2-function with bounded derivatives $(f \in C_b^2)$. By means of the Itô formula

$$f(X_t) = f(X_0) + \int_0^t f'(X_s) \, dB_s - \frac{1}{2} \int_0^t f'(X_s) X_s \, ds + \frac{1}{2} \int_0^t f''(X_s) \, ds,$$

we have

$$P_t f(x) = f(x) - \frac{1}{2} \int_0^t \mathbb{E}_x\big[f'(X_s) X_s\big] \, ds + \frac{1}{2} \int_0^t \mathbb{E}_x\big[f''(X_s)\big] \, ds.$$

From the bound

$$\int |x| \, dN(m, \sigma^2) \le \left(\int x^2 \, dN(m, \sigma^2) \right)^{1/2} = (m^2 + \sigma^2)^{1/2}$$

$$\mathbb{E}_x|X_s| \le (1 - e^{-s} + x^2 e^{-s})^{1/2} \le 1 + |x|$$

we observe that

$$\frac{1}{t}\big(P_t f(x) - f(x)\big) \xrightarrow[t \downarrow 0]{} \frac{1}{2} f''(x) - \frac{1}{2} x f'(x)$$

and

$$\frac{1}{t}\big(P_t f(x) - f(x)\big) - \frac{1}{2} f''(x) + \frac{1}{2} x f'(x)$$

remains bounded in absolute value by a function of the form $a|x| + b$ hence the convergence is in $L^2(m)$.

In other words, $f \in \mathcal{D}A$ and

$$Af(x) = \frac{1}{2}f''(x) - \frac{1}{2}xf'(x). \tag{10}$$

A fortiori $f \in \mathbb{D}$, and

$$\mathcal{E}[f] = \langle -Af, f\rangle_{L^2(m)} = -\int \left(\frac{1}{2}f''(x) - \frac{x}{2}f'(x)\right) f(x)\,dm(x).$$

Integration by parts yields

$$\mathcal{E}[f] = \frac{1}{2}\int f'^2(x)\,dm(x). \tag{11}$$

If we consider the bilinear operator Γ, defined by

$$\Gamma[f, g] = f'g', \tag{12}$$

and the associated quadratic operator

$$\Gamma[f] = f'^2,$$

we have

$$\mathcal{E}[f] = \frac{1}{2}\int \Gamma[f]\,dm$$

and for C_b^2 functions

$$\Gamma[f] = Af^2 - 2fAf. \tag{13}$$

Recapitulation

Let us emphasize the important properties obtained for the one-dimensional Ornstein–Uhlenbeck semigroup.

(P_t) is a symmetric strongly-continuous contraction semigroup on $L^2(m)$.

For $f \in L^2(m)$, $\mathcal{E}_t[f] = \frac{1}{t}\langle f - P_t f, f\rangle_{L^2(m)}$ is positive and increases as t tends to zero.

Defining $\mathbb{D} = \{f : \lim_{t\to 0} \mathcal{E}_t[f] < +\infty\}$ and for $f \in \mathbb{D}$

$$\mathcal{E}[f] = \lim_{t\to 0} \uparrow \mathcal{E}_t[f],$$

$\mathbb{D}$ is dense in $L^2(m)$, contains $\mathcal{D}A$ and is preserved by contractions:

F contraction and $f \in \mathbb{D}$ implies $F \circ f \in \mathbb{D}$ and

$$\mathcal{E}[F \circ f] \le \mathcal{E}[f].$$

The form $(\mathcal{E}, \mathbb{D})$ is closed.

A bilinear operator Γ exists, related to $\mathcal{E}$ by

$$\mathcal{E}[f] = \frac{1}{2} \int \Gamma[f] \, dm,$$

which satisfies for C^1 and Lipschitz functions F, G and $f \in \mathbb{D}^m$, $g \in \mathbb{D}^n$

$$\Gamma\big[F(f_1, \dots, f_m), G(g_1, \dots, g_m)\big] = \sum_{ij} F_i'(f) G_j'(g) \Gamma\big[f_i, g_j\big].$$

These properties will be axiomatized in the following chapter.

Comment. Let m be a positive σ-finite measure on the space E.

For a strongly-continuous contraction semigroup P_t on $L^2(E, m)$ symmetric with respect to m, the form $(\mathcal{E}, \mathbb{D})$ constructed as above is always closed.

For such a semigroup, the property that contractions operate on $\mathbb{D}$ and reduce $\mathcal{E}$ is equivalent to the property that P_t acts positively on positive functions. In this case P_t is sub-Markovian and gives rise (with the additional assumption of quasi-regularity) to a Markov process.

The theory of such semigroups and the associated Dirichlet forms were initiated by Beurling and Deny and then further developed by several authors, especially Fukushima.

Appendix. Determination of $\mathbb{D}$ for the Ornstein–Uhlenbeck semigroup

Let us introduce the Hermite polynomials

$$H_n(x) = e^{\frac{x^2}{2}} \left(-\frac{d}{dx}\right)^n \left(e^{-\frac{x^2}{2}}\right)$$

and their generating series

$$e^{xz - \frac{1}{2}z^2} = \sum_{n=0}^{\infty} \frac{z^n}{n!} H_n(x). \tag{14}$$

H_n is a polynomial of degree n, hence $H_n \in L^2(m)$.

Lemma II.10. $P_t H_n = e^{-\frac{tn}{2}} H_n$.

Proof. Setting $\xi_z(x) = e^{xz - \frac{1}{2}z^2}$, we directly compute $P_t \xi_z$ and obtain

$$P_t \xi_z = \xi_{\sqrt{e^{-t}}\, z}.$$

Using this along with (14) yields

$$\sum_n \frac{z^n}{n!} P_t H_n = \sum_n \frac{z^n}{n!} e^{-\frac{tn}{2}} H_n. \qquad \diamond$$

Lemma II.11. $\frac{H_n}{\sqrt{n!}}$ *is an orthonormal basis of* $L^2(m)$.

Proof. The fact that (H_n) is an orthogonal system is easy to proof and general for eigenvectors of symmetric operators. Taking the square of (14) and integrating with respect to m provides the norm of H_n. The system is complete in $L^2_{\mathbb{C}}(m)$ on the complex field $\mathbb{C}$, for the closed spanned space containing $x \to e^{xiu}$ by (14), hence functions fastly decreasing $\mathcal{S}$ which are dense in $L^2_{\mathbb{C}}(m)$. It is therefore complete in $L^2(m)$. $\diamond$

Let $f \in L^2(m)$ with expansion $f = \sum_n a_n \frac{H_n}{\sqrt{n!}}$. From

$$P_t f = \sum_n a_n e^{-\frac{tn}{2}} \frac{H_n}{\sqrt{n!}}$$

we can derive

$$\mathcal{E}_t[f] = \frac{1}{t}\langle f - P_t f, f\rangle = \sum_n \frac{1 - e^{-\frac{tn}{2}}}{t} a_n^2.$$

By the virtue of the simple fact that $\frac{1-e^{-\frac{tn}{2}}}{t} \uparrow \frac{n}{2}$ as $t \downarrow 0$, we obtain:

$$\mathbb{D} = \Big\{ f \in L^2 \colon \sum_n n a_n^2 < +\infty \Big\}$$

$$\mathcal{E}[f] = \sum_n \frac{n}{2} a_n^2 \quad \text{for } f \in \mathbb{D}.$$

Proposition II.12. $\mathbb{D} = \{ f \in L^2(m) \colon f' \textit{ in the distribution sense} \in L^2(m) \}$

$$\mathcal{E}[f] = \frac{1}{2} \| f' \|^2_{L^2(m)} \quad \textit{for } f \in \mathbb{D}.$$

Proof. We will use the two following elementary formulae:

$$H_{n+1}(x) = x H_n(x) - n H_{n-1}(x) \tag{15}$$

$$H_n'(x) = n H_{n-1}(x). \tag{16}$$

a) Let $f \in L^2(m)$, $f = \sum_n a_n \frac{H_n}{\sqrt{n!}}$ such that f' in the distribution sense belongs to $L^2(m)$.

The coefficients of the expansion of f' on the H_n's are given by

$$\int f' \frac{H_n}{\sqrt{n!}}\, dm = -\int f \left(\frac{H_n}{\sqrt{n!}} \frac{e^{-\frac{x^2}{2}}}{\sqrt{2\pi}} \right)' dx,$$

and thanks to (15)

$$= \int f \frac{H_{n+1}}{\sqrt{n!}}\, dm = a_{n+1}\sqrt{n+1}.$$

Hence $f' \in L^2(m)$ implies $\sum a_n^2 n < +\infty$.

b) Reciprocally if $f \in L^2(m)$, $f = \sum a_n \frac{H_n}{\sqrt{n!}}$ is such that $\sum_n a_n^2 n < +\infty$. Let g be the function

$$g = \sum_n \sqrt{n+1}\, a_{n+1} \frac{H_n}{\sqrt{n!}} \in L^2(m).$$

By (16)

$$\int_0^x g(y)\, dy = \sum_n \frac{a_{n+1}}{\sqrt{(n+1)!}} \big(H_{n+1}(x) - H_{n+1}(0)\big)$$

and by dominated convergence the series on the right-hand side converges for fixed x.

Now, the estimate $H_{2p+1}(0) = 0$, $H_{2p} = (-1)^p \frac{(2p)!}{2^p p!}$ shows that

$$\sum_n \frac{a_{n+1}}{\sqrt{(n+1)!}} H_{n+1}(x)$$

pointwise converges and coincides with the $L^2(m)$-expansion of f. Thus

$$\int_0^x g(y)\, dy = f(x) - f(0),$$

which proves the result. ◇

The same method shows that

$$\mathcal{D}A = \Big\{ f \in L^2(m) \colon \sum_n a_n^2 n^2 < +\infty \Big\}$$

$$Af = -\sum_n a_n \frac{n}{2} \frac{H_n}{\sqrt{n!}}$$

and similarly

Proposition II.13. $\mathcal{D}A = \big\{ f \in L^2(m) \colon (f'' - xf')$ *in the distribution sense belong to* $L^2(m) \big\}$ *and*

$$Af(x) = \frac{1}{2} f''(x) - \frac{1}{2} x f'(x).$$

Bibliography for Chapter II

N. Bouleau and F. Hirsch, *Dirichlet Forms and Analysis on Wiener Space*, Walter de Gruyter, 1991.

Cl. Dellacherie and P.-A. Meyer, *Probabilités et Potentiel*, Chap. XII à XVI, *Théorie du Potentiel associée à une résolvante, Théorie des processus de Markov*, Hermann 1987.

J. Deny, Méthodes hilbertiennes et théorie du potentiel, in: *Potential Theory*, C.I.M.E., Ed. Cremonese, Roma, 1970.

M. Fukushima, *Dirichlet Forms and Markov Processes*, North Holland, Kodansha, 1980.

M. Fukushima, Y. Oshima and M. Takeda, *Dirichlet Forms and Markov Processes*, Walter de Gruyter, 1994.

Z.-M. Ma and M. Röckner, *Introduction to the Theory of (Non-symmetric) Dirichlet Forms*, Springer-Verlag, 1992.

W. Rudin, *Functional Analysis*, McGraw-Hill, 1973.

G. E. Uhlenbeck and L. S. Ornstein, On the theory of Brownian motion, *Phys. Rev.* (2) **36** (1930), 823–841.

K. Yosida, *Functional Analysis*, Springer-Verlag, 1974.

Chapter III

Error structures

An error structure is a probability space equipped with a quadratic operator providing the variances and covariances of errors. In addition we impose that the associated quadratic form be closed.

This property plays a similar role in the theory as does the σ-additivity for probability spaces: it is an extension tool that allows passages to the limit under many useful circumstances.

The chapter begins by giving examples of error structures that illustrate the general definition. It will then be shown how computations can be performed in error structures. Attention will also focus on the existence of densities. Finally, sufficient conditions for closability will be studied.

1 Main definition and initial examples

Definition III.1. An *error structure* is a term

$$(\Omega, \mathcal{A}, \mathbb{P}, \mathbb{D}, \Gamma)$$

where $(\Omega, \mathcal{A}, \mathbb{P})$ is a probability space, and

(1) $\mathbb{D}$ is a dense subvector space of $L^2(\Omega, \mathcal{A}, \mathbb{P})$ (also denoted $L^2(\mathbb{P})$);

(2) Γ is a positive symmetric bilinear application from $\mathbb{D} \times \mathbb{D}$ into $L^1(\mathbb{P})$ satisfying "the functional calculus of class $\mathcal{C}^1 \cap \mathrm{Lip}$". This expression means

$$\forall u \in \mathbb{D}^m, \quad \forall v \in \mathbb{D}^n, \quad \forall F\colon \mathbb{R}^m \to \mathbb{R}, \quad \forall G\colon \mathbb{R}^n \to \mathbb{R}$$

with F, G being of class $\mathcal{C}^1$ and Lipschitzian, we have $F(u) \in \mathbb{D}$, $G(v) \in \mathbb{D}$ and

$$\Gamma[F(u), G(v)] = \sum_{i,j} \frac{\partial F}{\partial x_i}(u) \frac{\partial G}{\partial x_j}(v) \Gamma\big[u_i, v_j\big] \quad \mathbb{P}\text{-a.s.};$$

(3) the bilinear form $\mathcal{E}[u, v] = \frac{1}{2}\mathbb{E}\big[\Gamma[u, v]\big]$ is "closed". This means that the space $\mathbb{D}$ equipped with the norm

$$\|u\|_{\mathbb{D}} = \big(\|u\|^2_{L^2(\mathbb{P})} + \mathcal{E}[u, u]\big)^{1/2}$$

is complete.

If, in addition

(4) the constant function 1 belongs to $\mathbb{D}$ (which implies $\Gamma[1] = 0$ by property (2)),

we say that the error structure is *Markovian*.

We will always write $\mathcal{E}[u]$ for $\mathcal{E}[u, u]$ and $\Gamma[u]$ for $\Gamma[u, u]$.

Comments and links with the corresponding mathematical literature. First of all let us remark that by the functional calculus (property (2)) we express that the operator Γ satisfies formula (13) of Chapter I. In other words, we prolong the ideas of Gauss about error calculations.

The factor $\frac{1}{2}$ in the definition of the form $\mathcal{E}$ (property (3)) of course has no importance and is only convenient according to the usual notation used in the theory of symmetric semigroups. With this definition, the form $\mathcal{E}$ is known in the literature as a *local Dirichlet form* on $L^2(\Omega, \mathcal{A}, \mathbb{P})$ that possesses a "squared field" operator (or a "carré du champ" operator) Γ. These notions are usually studied on σ-finite measurable spaces. We limit ourselves herein to probability spaces both for the sake of simplicity and because we will often use images and products of error structures (see next chapter).

Under very weak additional assumptions (see Bouleau–Hirsch [1991], Ma–Röckner [1992]) to an error structure (also to a Dirichlet form on a σ-finite measurable space) a strongly-continuous contraction semigroup $(P_t)_{t\geq 0}$ on $L^2(\mathbb{P})$ can be uniquely associated, which is symmetric with respect to $\mathbb{P}$ and sub-Markov. This semigroup has a generator $(A, \mathcal{D}A)$, a self-adjoint operator that satisfies

$$(1) \qquad A\big[F(u)\big] = \sum_i \frac{\partial F}{\partial x_i}(u)A\big[u_i\big] + \frac{1}{2}\sum_{i,j} \frac{\partial^2 F}{\partial x_i \partial x_j}(u)\Gamma\big[u_i, u_j\big] \ \mathbb{P}\text{-a.s.}$$

for $F\colon \mathbb{R}^m \to \mathbb{R}$ of class $\mathcal{C}^2$ with bounded derivatives and $u \in (\mathcal{D}A)^m$ such that $\Gamma\big[u_i\big] \in L^2(\mathbb{P})$ (see Bouleau–Hirsch [1991]).

The Dirichlet form, the semigroup and the generator can also be made in correspondence with a resolvent family (see, for example Fukushima [1980]).

In order to clarify the intuitive meaning of Γ, we can suppose a larger σ-field $\mathcal{B}$ on Ω, $\mathcal{B} \supset \mathcal{A}$, such that the random variables and their errors be $\mathcal{B}$-measurable. Then for a random variable X, denoting its error by ΔX as in Chapter I, $\Gamma[X]$ must be considered as

$$\Gamma[X] = \lim \ \operatorname{var}[\Delta X \mid \mathcal{A}]$$

with the limit being taken as a normalization parameter that calibrates the size of the errors tends to zero. Similarly, the generator A describes the error biases:

$$A[X] = \lim \ \mathbb{E}[\Delta X \mid \mathcal{A}].$$

As discussed in Chapter I, error biases follow a second-order functional calculus (relation (1)), whereas error variances follow a first-order functional calculus (property (2) of the definition).

We will now illustrate the definition by means of some examples.

Example III.2.

$$\begin{aligned}\Omega &= \mathbb{R}\\ \mathcal{A} &= \text{Borel } \sigma\text{-field } \mathcal{B}(\mathbb{R})\\ \mathbb{P} &= \mathcal{N}(0,1) \text{ reduced normal law}\\ \mathbb{D} = H^1\big(\mathcal{N}(0,1)\big) &= \big\{u \in L^2(\mathbb{P}),\, u' \text{ in the distribution sense belongs to } L^2(\mathbb{P})\big\}\\ \Gamma[u] &= u'^2.\end{aligned}$$

Then, as a consequence of Chapter II, $\big(\mathbb{R}, \mathcal{B}(\mathbb{R}), \mathcal{N}(0,1), H^1(\mathcal{N}(0,1)), \Gamma\big)$ is an error structure. We also obtained the generator:

$$\mathcal{D}A = \big\{f \in L^2(\mathbb{P})\colon\ f'' - xf' \text{ in the distribution sense} \in L^2(\mathbb{P})\big\}$$

and

$$Af = \frac{1}{2}f'' - \frac{1}{2}I \cdot f'$$

where I is the identity map on $\mathbb{R}$.

Example III.3.

$$\begin{aligned}\Omega &= [0,1]\\ \mathcal{A} &= \text{Borel } \sigma\text{-field}\\ \mathbb{P} &= \text{Lebesgue measure}\\ \mathbb{D} &= \big\{u \in L^2\big([0,1], dx\big)\colon \text{the derivative } u' \text{ in the distribution}\\ &\qquad\qquad \text{sense over }]0,1[\text{ belongs to } L^2([0,1],dx)\big\}\\ \Gamma[u] &= u'^2.\end{aligned}$$

The space $\mathbb{D}$ defined herein is denoted $H^1\big([0,1]\big)$. Let us show that $(\Omega, \mathcal{A}, \mathbb{P}, \mathbb{D}, \Gamma)$ is an error structure.

1. $\mathbb{D}$ is dense in $L^2\big([0,1], dx\big)$ for $\mathbb{D} \supset \mathcal{C}_K^\infty\big(]0,1[\big)$.

2. It is known from the theory of distribution that if $v \in H^1\big([0,1]\big)$ (i.e., $v \in L^2$ and v' in the distribution sense $\in L^2$), then v possesses a continuous version which is derivable almost everywhere and whose derivative is a version of v'. (See Rudin [1966], Schwartz [1966].)

In addition if $u = (u_1, \dots, u_m) \in (H^1([0,1[))^m$ and if $F : \mathbb{R}^m \to \mathbb{R}$ is $\mathcal{C}^1 \cap \mathrm{Lip}$ then $F \circ u \in H^1(]0,1[)$ and

$$(F \circ u)' = \sum_i F_i' \circ u \cdot u_i'$$

in the sense of distribution hence almost everywhere. This shows that $\mathbb{D}$ is preserved by $\mathcal{C}^1 \cap \mathrm{Lip}$-functions and that Γ satisfies the functional calculus of class $\mathcal{C}^1 \cap \mathrm{Lip}$.

3. To show that the form $\mathcal{E}[u,v] = \frac{1}{2}\mathbb{E}\Gamma[u,v]$ is closed, let us put

$$\|u\|_{\mathbb{D}} = \left(\mathcal{E}[u] + \|u\|_{L^2}^2\right)^{1/2}$$

and let u_n be a $\|\cdot\|_{\mathbb{D}}$-Cauchy sequence.

There exists $u, f \in L^2$ such that

$$\begin{aligned} u_n &\to u \quad \text{in } L^2 \\ u_n' &\to f \quad \text{in } L^2. \end{aligned}$$

Let $\varphi \in \mathcal{C}_K^\infty(]0,1[)$, we have

$$\begin{aligned} \int_0^1 \varphi(x) f(x)\,dx &= \lim_{n\uparrow\infty} \int_0^1 \varphi(x) u_n'(x)\,dx \\ &= \lim_{n\uparrow\infty} -\int_0^1 \varphi'(x) u_n(x) = -\int_0^1 \varphi' u\,dx. \end{aligned}$$

Hence f is the derivative of u in the distribution sense, i.e. $u \in \mathbb{D}$ and $u_n \to u$ in $\mathbb{D}$.

Remark III.4. In this example, the convergence in $\mathbb{D}$ preserves continuity (to be more formal, the existence of a continuous version is actually preserved). This stems from both equicontinuity and the Ascoli theorem. Indeed let u_n be a Cauchy sequence in $\mathbb{D}$ and $\tilde{u}_n$ be continuous versions of the u_n's. We then have

$$\left|\tilde{u}_n(y) - \tilde{u}_n(x)\right|^2 = \left|\int_x^y u_n'(t)\,dt\right|^2 \le |y-x| \int_0^1 u_n'^2(t)\,dt,$$

but the $\left\|u_n'\right\|_{L^2}$ are bounded (a Cauchy sequence is bounded): the $\tilde{u}_n$'s are equi-uniformly-continuous on $[0,1]$.

According to the Ascoli theorem (see Rudin [1966]) a uniformly-converging subsequence exists, such that the limit of the u_n's possesses a continuous version.

Remark III.5. In order to identify the generator of this error structure, let us admit the following lemma from the theory of symmetric semigroups.

Lemma. *Let $f \in \mathbb{D}$. Then $f \in \mathcal{D}A$ if and only if there exists $g \in L^2(\mathbb{P})$ such that*

$$\mathcal{E}[f,u] = \langle g,u\rangle_{L^2(\mathbb{P})} \quad \forall u \in \mathbb{D}_0$$

where $\mathbb{D}_0$ is a dense subset in $\mathbb{D}$. If this condition is fulfilled then $Af = -g$.

Hence, in our case

$$-\int_0^1 Af(x)g(x)\,dx = \frac{1}{2}\int_0^1 f'(x)g'(x)\,dx \quad \forall g \in \mathcal{C}^1([0,1]).$$

On this equation, we can observe by means of integration by parts in the second term, that

$$\mathcal{D}A \supset \left\{f \in \mathcal{C}^2([0,1]) \colon f'(0) = f'(1) = 0\right\}$$

and for such a function f,

$$Af = \frac{1}{2}f''.$$

Example III.6. Let U be a domain (connected open set) in $\mathbb{R}^d$ with unit volume, $\mathcal{B}(U)$ be the Borel σ-field and $dx = dx_1, \dots dx_d$ be the Lebesgue measure

$$\mathbb{D} = \big\{u \in L^2(U,dx) \colon \text{the gradient } \nabla u \text{ in the distribution sense belongs to } L^2\big(U,dx;\mathbb{R}^d\big)\big\}$$

$$\Gamma[u] = |\nabla u|^2 = \left(\frac{\partial u}{\partial x_1}\right)^2 + \dots + \left(\frac{\partial u}{\partial x_d}\right)^2.$$

Then $(U, \mathcal{B}(U), dx, \mathbb{D}, \Gamma)$ is an error structure.

Proof. 1. $\mathbb{D}$ is dense in $L^2(U,dx)$ since $\mathcal{C}_K^\infty(U) \subset \mathbb{D}$, and $\mathcal{C}_K^\infty(U)$ is dense.

2. We will admit the following lemma from the theory of distributions.

Lemma. *Let $w \in L^2(U,dx)$ be such that $\forall i = 1,\dots,d$ $\frac{\partial w}{\partial x_i}$ in the distribution sense belongs to $L^2(U,dx)$. Then for $G \in \mathcal{C}^1 \cap \mathrm{Lip}$, $G \circ w \in L^2(U,dx)$, $\frac{\partial G \circ w}{\partial x_i}$ in the distribution sense belongs to $L^2(U,dx)$ and*

$$\frac{\partial G \circ w}{\partial x_i} = G' \circ w \cdot \frac{\partial w}{\partial x_i}.$$

According to the lemma, if $v = (v_1,\dots,v_m) \in \mathbb{D}^m$ and if $F \colon \mathbb{R}^m \to \mathbb{R}$ is $\mathcal{C}^1 \cap \mathrm{Lip}$, the gradient

$$\nabla(F \circ v) = \left(F_1' \circ v\frac{\partial v_1}{\partial x_1} + \dots + F_m' \circ v\frac{\partial v_m}{\partial x_1}, \dots, F_1' \circ v\frac{\partial v_1}{\partial x_d} + \dots + F_m' \circ v\frac{\partial v_m}{\partial x_d}\right)$$

belongs to $L^2(U,dx;\mathbb{R}^d)$ and the formula of the functional calculus for Γ follows.

3. To show that the form $\mathcal{E}$ associated with Γ is closed, we will proceed as in Example III.3.

Let u_n be a Cauchy sequence in $(\mathbb{D}, \|\cdot\|_{\mathbb{D}})$. There exists $u \in L^2(U, dx)$ and $f = (f_1, \dots, f_d) \in L^2(U, dx; \mathbb{R})$ such that

$$\begin{aligned} u_n &\to u \quad \text{in } L^2(U, dx) \\ \nabla u_n &\to f \quad \text{in } L\big(U, dx; \mathbb{R}^d\big). \end{aligned}$$

If $\varphi \in \mathcal{C}_K^\infty(U)$ we have

$$\begin{aligned} \int_{\mathbb{R}^d} f_i \varphi \, dx = \lim_n \int \frac{\partial u_n}{\partial x_i} \varphi \, dx &= -\lim_n \int u_n \frac{\partial \varphi}{\partial x_i} \, dx \\ &= -\int u \frac{\partial \varphi}{\partial x_i} \, dx, \end{aligned}$$

hence $f_i = \frac{\partial u}{\partial x_i}$ in the distribution sense. Thus, $u \in \mathbb{D}$ and $u_n \to u$ in $\mathbb{D}$.

4. Assumption 4 is satisfied, this error structure is therefore Markovian. ◇

Remark. From the relation $\mathcal{E}[f, g] = \langle -Af, g \rangle$ we see easily that the domain of the generator contains the functions of class $\mathcal{C}^2$ with compact support in U, $\mathcal{D}A \supset \mathcal{C}_K^2(U)$ and that for such functions

$$Af = \frac{1}{2} \Delta f = \frac{1}{2} \sum_{i=1}^d \frac{\partial^2 f}{\partial x_i^2}.$$

2 Performing calculations in error structures

Let us mention three easy facts stemming from the definition. Here $(\Omega, \mathcal{A}, \mathbb{P}, \mathbb{D}, \Gamma)$ is an error structure and $\mathbb{D}$ is always equipped with the norm $\|\cdot\|_{\mathbb{D}}$ defined in Point 3.

2.1. The positivity of Γ ($\Gamma[u] \geq 0 \; \forall u \in \mathbb{D}$) implies that

$$\big|\Gamma[u, v]\big| \leq \sqrt{\Gamma[u]} \sqrt{\Gamma[v]}, \qquad u, v \in \mathbb{D},$$

and

$$\mathbb{E}\big|\Gamma[u, v] - \Gamma[u_1, v_1]\big| \leq \sqrt{\mathbb{E}\big[\Gamma[u - u_1]\big]} \sqrt{\mathbb{E}\Gamma[v]} + \sqrt{\mathbb{E}\Gamma[u_1]} \sqrt{\mathbb{E}\Gamma[v - v_1]}.$$

We see that Γ *is continuous from* $\mathbb{D} \times \mathbb{D}$ *into* $L^1(\mathbb{P})$.

2.2. *If $u \in \mathbb{D}$, the sequence of bounded functions $u_n = n \operatorname{Arctan} \frac{u}{n}$ converges to u, $\mathbb{P}$-a.e. and in $\mathbb{D}$.*

Indeed $u_n \to u$ $\mathbb{P}$-a.e. and in $L^2(\mathbb{P})$ and according to the functional calculus

$$\Gamma\big[u - u_n\big] = \left(1 - \frac{1}{1 + \frac{u^2}{n^2}}\right)^2 \Gamma[u]$$

tends to zero in $L^1(\mathbb{P})$.

2.3. *If the sequence u_n converges to u in $\mathbb{D}$, there exists a subsequence u_{n_k} converging to u $\mathbb{P}$-a.e. and in $\mathbb{D}$.*

The following property is often useful in order to prove that a given function is in $\mathbb{D}$.

2.4. *If the sequence u_n is weakly-bounded in $\mathbb{D}$ and converges to u in $L^2(\mathbb{P})$, then $u \in \mathbb{D}$.*

Proof. The hypothesis states: $u_n \in \mathbb{D}$ and the sequence $\langle u_n, v\rangle_{\mathbb{D}}$ is bounded $\forall v \in \mathbb{D}$. This implies that (u_n) is strongly-bounded and hence weakly relatively compact and there are $w \in \mathbb{D}$ and a subsequence u_{n_k} such that $u_{n_k} \to w$ weakly in $\mathbb{D}$.

It then follows that there is a convex combination of the u_{n_k} strongly converging to w. Hence necessarily $u = w$. ◇

2.5. Assumption (2) of the definition of an error structure may be weakened. If we change it to

(2*) Γ is a positive symmetric bilinear mapping from $\mathbb{D} \times \mathbb{D}$ into $L^1(\mathbb{P})$ such that there exists $\mathbb{D}_0$ dense in $\mathbb{D}$ (for the norm $\|\cdot\|_{\mathbb{D}}$) such that if $u \in \mathbb{D}_0^m$, $v \in \mathbb{D}_0^n$, $F \in \mathcal{C}^1 \cap \operatorname{Lip}(\mathbb{R}^m)$, $G \in \mathcal{C}^1 \cap \operatorname{Lip}(\mathbb{R}^n)$, then $F \circ u \in \mathbb{D}$, $G \circ v \in \mathbb{D}$ and

$$\Gamma[F \circ u, G \circ v] = \sum_{ij} \frac{\partial F}{\partial x_i} \circ u \cdot \frac{\partial G}{\partial x_j} \circ v \cdot \Gamma\big[u_i, v_j\big] \ \mathbb{P}\text{-a.e.},$$

holding the other assumptions unchanged, then Assumption (2) is fulfilled.

Proof. a) Let $u \in \mathbb{D}$ and $F \in \mathcal{C}^1 \cap \operatorname{Lip}(\mathbb{R})$. Consider a sequence $u_n \in \mathbb{D}_0$ such that $u_n \to u$ $\mathbb{P}$-a.e. and in $\mathbb{D}$.

We first have

$$\|F(u_n) - F(u)\|_{L^2(\mathbb{P})} \le K\|u_n - u\|_{L^2(\mathbb{P})} \xrightarrow[n\uparrow\infty]{} 0$$

where K is the Lipschitz constant of F. Then

$$\begin{aligned}
2\mathcal{E}\big[F(u_p) - F(u_q)\big] & \\
&= \mathbb{E}\big\{\Gamma[F(u_p)] - 2\Gamma[F(u_p), F(u_q)] + \Gamma[(u_q)]\big\} \\
&= \mathbb{E}\big\{F'^2(u_p)\Gamma[u_p] - 2F'(u_p)F'(u_q)\Gamma[u_p, u_q] + F'^2(u_q)\Gamma[u_q]\big\}.
\end{aligned}$$

From the continuity of Γ (Argument 2.1 does not use functional calculus), the quantities $\Gamma\big[u_p\big]$, $\Gamma\big[u_p, u_q\big]$, $\Gamma\big[u_q\big]$ tend to $\Gamma[u]$ in $L^1(\mathbb{P})$ as $p, q \uparrow \infty$. By dominated convergence $\mathcal{E}\big[F(u_p) - F(u_q)\big] \to 0$ as $p, q \uparrow \infty$. Now, the form $\mathcal{E}$ is closed, so $F(u) \in \mathbb{D}$ and $F\big(u_n\big) \to F(u)$ in $\mathbb{D}$.

b) Let $u \in \mathbb{D}^m$, $v \in \mathbb{D}^n$, $F \in \mathcal{C}^1 \cap \mathrm{Lip}\big(\mathbb{R}^m\big)$, $G \in \mathcal{C}^1 \cap \mathrm{Lip}\big(\mathbb{R}^n\big)$. Consider sequences $u_k \in \mathbb{D}_0^m$ and $v_k \in \mathbb{D}_0^n$ such that $u_k \to u$ $\mathbb{P}$-a.e. and in $\mathbb{D}^m$ and $v_k \to 0$ $\mathbb{P}$-a.e. and in $\mathbb{D}^n$. Using the same argument as in a), $F \circ u \in \mathbb{D}$, $G \circ v \in \mathbb{D}$ and $F \circ u_k \to F \circ u$ in $\mathbb{D}$ and $G \circ v_k \to G \circ v$ in $\mathbb{D}$. In the equality

$$\Gamma\big[F \circ u_k, G \circ u_k\big] = \sum_{ij} F_i' \circ u_k \cdot G_j' \circ v_k \Gamma\big[u_{k,i}, v_{k,j}\big],$$

the left-hand side tends to $\Gamma[F \circ u, G \circ v]$ in $L^1(\mathbb{P})$, by virtue of the continuity of Γ, and the right-hand side tends to

$$\sum_{ij} F_i' \circ u\, G_j' \circ v\, \Gamma\big[u_i, v_j\big]$$

in $L^1(\mathbb{P})$ by continuity of Γ and by the continuity and boundedness of the derivative of F and G. ◇

The vector space $\mathbb{D}$ is preserved not only by $\mathcal{C}^1 \cap \mathrm{Lip}$ functions but by Lipschitz functions.

To prove this we use the following lemma.

Lemma III.7. *Let μ be a probability measure on $\mathbb{R}^m$, let $|\cdot|$ be one of the equivalent norms on $\mathbb{R}^n$, and let F be a Lipschitz function on $\mathbb{R}^m$ with constant K for the norm $|\cdot|$.*

a) *There exist functions $F_k \in \mathcal{C}^\infty \cap \mathrm{Lip}\big(\mathbb{R}^m\big)$ with same Lipschitz constant K as F, such that $F_k \xrightarrow[k\uparrow\infty]{} F$ everywhere on $\mathbb{R}^m$ and such that the derivatives $\frac{\partial F_k}{\partial x_i}$ converge μ-a.e.*

b) *If F is $\mathcal{C}^1$, the F_k can be chosen such that in addition*

$$\forall i = 1, \dots, m, \quad \frac{\partial F_k}{\partial x_i} \xrightarrow[k\uparrow\infty]{} \frac{\partial F}{\partial x_i} \quad \textit{everywhere.}$$

Proof. Let $\alpha_k \in \mathcal{C}_K^\infty(\mathbb{R}^m)$, $\alpha_k \geq 0$, $\int \alpha_k\,dx = 1$, such that the support of $\alpha_k \xrightarrow[k\uparrow\infty]{} \{0\}$. Let us set

$$F_k = F * \alpha_k.$$

The F_k are Lipschitz with constant K and $F_k \xrightarrow[k\uparrow\infty]{} F$ everywhere.

If F is $\mathcal{C}^1$, $\frac{\partial F_k}{\partial x_i} \xrightarrow[k\uparrow\infty]{} \frac{\partial F}{\partial x_i}$ everywhere, thereby proving b).

If F is only Lipschitz, the functions $\frac{\partial F_k}{\partial x_i}$ satisfy $\left|\frac{\partial F_k}{\partial x_i}\right| \leq K$ and the bounded sequence $\left(\frac{\partial F_k}{\partial x_i}\right)_{k\geq 0}$ is relatively compact in $L^2(\mu)$ for the weak topology. A function ψ_i exists such that a subsequence $\frac{\partial F_{k'}}{\partial x_i}$ weakly converges to ψ_i. By means of a classical result, there are convex combinations, which are derivatives of the same convex combinations of the F_k, which converge to ψ_i in $L^2(\mu)$. Then, by once again extracting a subsequence, we obtain a family converging $\mathbb{P}$-a.e. and satisfying the statement of the lemma. ◇

Proposition III.8. *Let $F\colon \mathbb{R}^m \to \mathbb{R}$ be Lipschitz, and let $u \in \mathbb{D}^m$, then $F \circ u \in \mathbb{D}$.*

Proof. By virtue of the lemma, there are approximations $F_k \in \mathcal{C}^\infty \cap \mathrm{Lip}$ with the same Lipschitz constant K as F, such that $F_k \to F$ everywhere and $\frac{\partial F_k}{\partial x_i}$ converges almost surely for the law of u. Then $F_k \circ u$ is Cauchy in $\mathbb{D}$. Indeed, $F_k \circ u \to F \circ u$ remaining dominated in absolute value, for large k, by $1 + |F(0)| + K|u|$, hence $F_k \circ u \xrightarrow[h\uparrow\infty]{} F \circ u$ in $L^2(\mathbb{P})$. Moreover,

$$2\mathcal{E}\big[F_k\circ u - F_{k'}\circ u\big] = \int \sum_{ij}\left(\frac{\partial F_k}{\partial x_i}\circ u - \frac{\partial F_{k'}}{\partial x_i}\circ u\right)\left(\frac{\partial F_h}{\partial x_j}\circ u - \frac{\partial F_h}{\partial x_j}\circ u\right)\Gamma\big[u_i, u_j\big]\,d\mathbb{P}$$

tends to zero by dominated convergence. From the closedness of $\mathcal{E}$, we get $F \circ u \in \mathbb{D}$ and $F_k \circ u \to F \circ u$ in $\mathbb{D}$, hence we also obtain

$$\mathcal{E}[F\circ u] = \lim_k \mathcal{E}\big[F_k \circ u\big] \leq K^2\mathcal{E}[u]. \qquad \diamond$$

Proposition III.9. *If F is a contraction, i.e.*

$$|F(x) - F(y)| \leq \sum_{i=1}^m \big|x_i - y_i\big|,$$

then for $u \in \mathbb{D}^m$ we have

$$\big(\Gamma[F\circ u]\big)^{1/2} \leq \sum_i \big(\Gamma[u_i]\big)^{1/2}$$

and

$$\big(\mathcal{E}[F\circ u]\big)^{1/2} \leq \sum_i \big(\mathcal{E}[u_i]\big)^{1/2}.$$

Proof. F is Lipschitz with constant 1 for the norm $|x| = \sum_{i=1}^m |x_i|$ on $\mathbb{R}^m$. Let $F_k \in \mathcal{C}^1 \cap \mathrm{Lip}$ with the same Lipschitz constant, such that $F_k \to F$ everywhere and such that $\frac{\partial F_k}{\partial x_i}$ converges almost surely for the law of u. We then know (cf. proof of Prop III.8) that $F_k \circ u \to F \circ u$ in $\mathbb{D}$. From the equality

$$\Gamma[F_k \circ u] = \sum_{ij} \frac{\partial F_k}{\partial x_i} \circ u \cdot \frac{\partial F_k}{\partial x_j} \circ u \cdot \Gamma[u_i, u_j]$$

and from

$$|\Gamma[u_i, u_j]| \leq \sqrt{\Gamma[u_i]}\sqrt{\Gamma[u_j]}$$

we derive

$$\Gamma[F_k \circ u] \leq \Big(\sum_i \Big|\frac{\partial F_k}{\partial x_i} \circ u\Big| \sqrt{\Gamma[u_i]}\Big)^2 \leq \Big(\sum_i \sqrt{\Gamma[u_i]}\Big)^2,$$

which yields the inequality for Γ by passage to the limit and the continuity of Γ.

The second inequality easily follows. ◇

Proposition III.10. *$\mathbb{D} \cap L^\infty(\mathbb{P})$ is an algebra, dense in $\mathbb{D}$. If $u, v \in \mathbb{D} \cap L^\infty$,*

$$(\mathcal{E}[u, v])^{1/2} \leq (\mathcal{E}[u])^{1/2}\|v\|_\infty + (\mathcal{E}[v])^{1/2}\|u\|_\infty.$$

Proof. If $f \in \mathbb{D}$, then n Arctan $\frac{f}{n}$ belongs to $\mathbb{D} \cap L^\infty$ and converges to f in $\mathbb{D}$ (see 2.2, p. 38), hence $\mathbb{D} \cap L^\infty$ is dense. The remainder of the argument proceeds exactly as for the Ornstein–Uhlenbeck structure on $\mathbb{R}$ (see Chapter II). ◇

3 Lipschitz functional calculus and existence of densities

Let $(\Omega, \mathcal{A}, \mathbb{P}, \mathbb{D}, \Gamma)$ be an error structure. The existence of the operator Γ in addition to the probability space $(\Omega, \mathcal{A}, \mathbb{P})$ allows to express sufficient conditions for probabilistic properties such as the existence of densities.

This kind of argument has received a considerable mathematical extension.

Let $u \in \mathbb{D}$ and μ the probability measure on $\mathbb{R}$ which is the law of u. Let g be Borel on $\mathbb{R}$, $|g| \leq 1$. We set $G(x) = \int_0^x g(t)\,dt$.

Lemma III.11. *There exists a sequence (g_n) of continuous functions on $\mathbb{R}$, $|g_n| \leq 1$, such that*

$$g_n \to g \ \ (dx + \mu)\text{-a.e.}$$

Proof. In this instance dx is the Lebesgue measure on $\mathbb{R}$. The lemma easily follows from the fact that continuous functions with compact support are dense in $L^1(dx+\mu)$. ◇

Let us define

$$G_n(x) = \int_0^x g_n(t)\,dt.$$

From the functional calculus of class $\mathcal{C}^1 \cap \mathrm{Lip}$, we have

$$\Gamma\big[G_n \circ u\big] = g_n^2 \circ u \,\Gamma[u]$$

and

$$\mathcal{E}\big[G_n \circ u - G_m \circ u\big] = \frac{1}{2}\int \big(g_n - g_m\big)^2 \circ u \cdot \Gamma[u]\,d\mathbb{P}.$$

This expression, coupled with the fact that $G_n \circ u$ tends to $G \circ u$ in $L^2(\mathbb{P})$ by dominated convergence, implies that the sequence $G_n \circ u$ is Cauchy in $\mathbb{D}$, and, therefore, converges to $G \circ u$ in $\mathbb{D}$. From the continuity of Γ

$$\Gamma[G \circ u] = \lim_n g_n^2 \circ u \,\Gamma[u] \;\text{ in } L^1(\mathbb{P})$$

however $g_n \circ u \xrightarrow[n\uparrow\infty]{} g \circ u$ $\mathbb{P}$-a.s. (because $g_n \to g$ μ-a.s.), hence

$$\Gamma[G \circ u] = g^2 \circ u \cdot \Gamma[u] \quad \mathbb{P}\text{-a.s.}$$

We then obtain

Theorem III.12. *For all $u \in \mathbb{D}$, the image by u of the (positive bounded) measure $\Gamma[u] \cdot \mathbb{P}$ is absolutely continuous with respect to the Lebesgue measure on $\mathbb{R}$:*

$$u_*\big(\Gamma[u] \cdot \mathbb{P}\big) \ll dx.$$

If $F\colon \mathbb{R} \to \mathbb{R}$ is Lipschitz

$$\Gamma[F \circ u] = F'^2 \circ u \cdot \Gamma[u]$$

where F' is any version of the derivative (defined Lebesgue-a.e.) of F.

Proof. Taking $g = 1_A$ where A is Lebesgue negligible yields the theorem. ◇

Thanks to this theorem, the operator Γ can be extended to a larger space than $\mathbb{D}$:

Definition III.13. A function $u\colon \Omega \to \mathbb{R}$ is said to be *locally in* $\mathbb{D}$, and we write $u \in \mathbb{D}_{\mathrm{loc}}$, if a sequence of sets $\Omega_n \in \mathcal{A}$ exists such that

- $\bigcup_n \Omega_n = \Omega$
- $\forall n\ \exists u_n \in \mathbb{D}\colon u_n = u$ on Ω_n.

$\mathbb{D}_{\text{loc}}$ is preserved by locally Lipschitz functions.

Proposition III.14. *Let u be in $\mathbb{D}_{\text{loc}}$.*

1) *There exists a unique positive class $\Gamma[u]$ (defined $\mathbb{P}$-a.e.) such that*

$$\forall v \in \mathbb{D}, \ \ \forall B \in \mathcal{A}, \ \ u = v \text{ on } B \Rightarrow \Gamma[u] = \Gamma[v] \ \text{ on } B.$$

2) *The image by u of the σ-finite measure $\Gamma[u] \cdot \mathbb{P}$ is absolutely continuous with respect to the Lebesgue measure.*

3) *If $F : \mathbb{R} \to \mathbb{R}$ is locally Lipschitz, $F \circ u \in \mathbb{D}_{\text{loc}}$ and*

$$\Gamma[F \circ u] = F'^2 \circ u \cdot \Gamma[u].$$

Proof. Let (u_n) and (Ω_n) be the localizing sequences for $u \in \mathbb{D}_{\text{loc}}$ (Definition III.13). It is then possible to define $\Gamma[u] = \Gamma\big[u_n\big]$ on Ω_n.

Indeed, suppose $v \in \mathbb{D}$ and $w \in \mathbb{D}$ coincide with u on $B \in \mathcal{A}$. We have

$$B \subset (v - w)^{-1}\big(\{0\}\big)$$

and according to Theorem III.12, B is negligible for the measure $\Gamma[v - w] \cdot \mathbb{P}$ which implies

$$\Gamma[v - w, v + w] = 0 \ \ \mathbb{P}\text{-a.s. on } B$$

and

$$\Gamma[v] = \Gamma[w] \quad \text{on } B.$$

This demonstration proves the first point and the others easily follow. ◇

We have observed (Proposition III.8) that $\mathbb{D}$ is stable by Lipschitz functions of several variables. We can extend the functional calculus to Lipschitz functions with an additional hypothesis.

Proposition III.15. *Let $u \in \mathbb{D}^m$, $v \in \mathbb{D}^n$ and let $F : \mathbb{R}^m \to \mathbb{R}$ and $G : \mathbb{R}^n \to \mathbb{R}$ be Lipschitz. Suppose the law of u is absolutely continuous with respect to the Lebesgue measure on $\mathbb{R}^m$ and the same for v on $\mathbb{R}^n$. Then*

$$\Gamma[F \circ u, G \circ v] = \sum_{i,j} \frac{\partial F}{\partial x_i} \circ u \, \frac{\partial G}{\partial x_j} \circ v \, \Gamma\big[u_i, v_j\big].$$

Proof. It is known that Lipschitz functions possess derivatives Lebesgue-a.e. Now if F is Lipschitz there exists an $\tilde{F}$ of class $\mathcal{C}^1 \cap \mathrm{Lip}$ that coincides with F outside a set of small Lebesgue measure. The same applies for G (see Morgan [1988], or Mattila [1995]).

The functional calculus applied to $\tilde{F}$ and $\tilde{G}$ and the fact that $\Gamma[F \circ u, G \circ v]$ coincides with $\Gamma[\tilde{F} \circ u, \tilde{G} \circ v]$ outside a set of small $\mathbb{P}$-measure by means of Proposition III.14 yields the result. ◇

The extension of Theorem III.12 to the case of $u = (u_1, \ldots, u_m) \in \mathbb{D}^m$ and to Lipschitz $F\colon \mathbb{R}^m \to \mathbb{R}$ has remained up until now conjecture. Nevertheless, the following result has been demonstrated for special cases including the classical case on $\mathbb{R}^d$ or $[0, 1]^d$ with $\Gamma = |\nabla|^2$ and that of Wiener space equipped with the Ornstein–Uhlenbeck form (see the following chapters). It is a useful tool for obtaining the existence of densities for random variables encountered in stochastic analysis, e.g. solutions to stochastic differential equations (see Bouleau–Hirsch [1991]).

Proposition III.16 (proved for special error structures). *If $u = (u_1, \ldots, u_m) \in \mathbb{D}^m$, then the image by u of the measure* $\det \Gamma[u_i, u_j]$. $\mathbb{P}$ *is absolutely continuous with respect to the Lebesgue measure on $\mathbb{R}^m$.*

On the other hand, no error structure is known at present that does not satisfy Proposition III.16.

4 Closability of pre-structures and other examples

It often arises that the domain $\mathbb{D}$ is not completely known and that only sufficient conditions are available for belonging to $\mathbb{D}$. We thus have to express the closedness of the form $\mathcal{E}$ using only a subspace of $\mathbb{D}$.

Definition III.17. Let $(\Omega, \mathcal{A}, \mathbb{P})$ be a probability space and $\mathbb{D}_0$ be a subvector space of $L^2(\mathbb{P})$. A positive symmetric bilinear form Q defined on $\mathbb{D}_0 \times \mathbb{D}_0$ is said to be *closable* if any Q-Cauchy sequence in $\mathbb{D}_0$ converging to zero in $L^2(\mathbb{P})$ converges to zero for Q:

$$u_n \in \mathbb{D}_0, \ \|u_n\|_{L^2(\mathbb{P})} \longrightarrow 0, \ Q[u_n - u_m] \xrightarrow[m,n\uparrow\infty]{} 0 \quad \text{implies} \quad Q[u_n] \xrightarrow[n\uparrow\infty]{} 0.$$

If Q is closable, it possesses a smallest closed extension. We can sketch out the standard mathematical procedure: Let $\mathcal{J}$ be the set of Q-Cauchy sequences in $\mathbb{D}_0$. On $\mathbb{D}_0$ we set the norm $N[\cdot] = \left(\|\cdot\|^2_{L^2(\mathbb{P})} + Q[\cdot]\right)^{1/2}$. The relation $\mathcal{R}$ on $\mathcal{J}$ defined by $(u_n)\mathcal{R}(v_n)$ if and only if $N[u_n - v_n] \xrightarrow[n\uparrow\infty]{} 0$ is an equivalence relation on $\mathcal{J}$. If $\mathbb{D}$ is defined as $\mathcal{J}/\mathcal{R}$, the elements of $\mathbb{D}$ can be identified with functions in $L^2(\mathbb{P})$ and Q extends to $\mathbb{D}$ as a closed form $\overline{Q}$.

Definition III.18. A term $(\Omega, \mathcal{A}, \mathbb{P}, \mathbb{D}_0, \Gamma)$, where $(\Omega, \mathcal{A}, \mathbb{P})$ is a probability space, is called an *error pre-structure* if

1) $\mathbb{D}_0$ is a dense subvector space of $L^2(\mathbb{P})$.

2) Γ is a positive symmetric operator from $\mathbb{D}_0 \times \mathbb{D}_0$ into $L^1(\mathbb{P})$ that satisfies the functional calculus of class $\mathcal{C}^\infty \cap \text{Lip}$ on $\mathbb{D}_0$. This means the following:

$$\forall u \in \mathbb{D}_0^m \quad \forall v \in \mathbb{D}_0^n, \quad \forall F\colon \mathbb{R}^m \to \mathbb{R}, \ \forall G\colon \mathbb{R}^n \to \mathbb{R},$$

F, G of class $\mathcal{C}^\infty$ and Lipschitz, then $F \circ u$ and $G \circ v$ are in $\mathbb{D}_0$ and

$$\Gamma[F \circ u, G \circ v] = \sum_{ij} F_i' \circ u\, G_j' \circ v\, \Gamma\big[u_i, v_j\big].$$

We can now prove that a pre-structure with a closable form extends to an error structure.

Proposition III.19. *Let* $(\Omega, \mathcal{A}, \mathbb{P}, \mathbb{D}_0, \Gamma)$ *be an error pre-structure such that the form* $\mathcal{E}[\cdot] = \frac{1}{2}\mathbb{E}\big[\Gamma[\cdot]\big]$ *defined on* $\mathbb{D}_0$ *is closable. Let* $\mathbb{D}$ *be the domain of the smallest closed extension of* $\mathcal{E}$, *then* Γ *extends to* $\mathbb{D}$ *and* $(\Omega, \mathcal{A}, \mathbb{P}, \mathbb{D}, \Gamma)$ *is an error structure.*

Proof. a) Let us denote $(\mathbb{D}, \mathcal{E})$ the smallest closed extension of $\big(\mathbb{D}_0, \mathcal{E}\big)$ and let us set

$$\|\cdot\|_{\mathbb{D}} = \big(\|\cdot\|^2_{L^2(\mathbb{P})} + \mathcal{E}[\cdot]\big)^{1/2}.$$

If $u_n \in \mathbb{D}_0$ converge to $u \in \mathbb{D}$, the inequality

$$\begin{aligned}\big|\Gamma[u_n] - \Gamma[u_m]\big| &= \big|\Gamma[u_n, u_n - u_m] + \Gamma[u_n - u_m, u_m]\big| \\ &\le \Big(\sqrt{\Gamma[u_m]} + \sqrt{\Gamma[u_n]}\Big)\sqrt{\Gamma[u_m - u_n]}\end{aligned}$$

yields

$$\frac{1}{2}\mathbb{E}\big|\Gamma[u_n] - \Gamma[u_m]\big| \le \Big(\sqrt{\mathcal{E}[u_m]} + \sqrt{\mathcal{E}[u_n]}\Big)\sqrt{\mathcal{E}[u_m - u_n]}$$

and shows that $\Gamma[u_n]$ converges in $L^1(\mathbb{P})$ to a value not depending on the sequence (u_n), but only on u, which will be denoted $\Gamma[u]$. The extension by bilinearity of Γ to $\mathbb{D} \times \mathbb{D}$ satisfies

$$\frac{1}{2}\mathbb{E}\Gamma[u, v] = \mathcal{E}[u, v] \quad \forall u, v \in \mathbb{D}$$

and, from the argument of 2.1, p. 37, Γ is continuous from $\mathbb{D} \times \mathbb{D}$ into $L^1(\mathbb{P})$.

b) The functional calculus of class $\mathcal{C}^\infty \cap \text{Lip}$ extends from functions in $\mathbb{D}_0$ to functions in $\mathbb{D}$ with the same argument used for Condition (2*) in Section 2.5.

c) It remains to extend the functional calculus of class $\mathcal{C}^\infty \cap \text{Lip}$ to a functional calculus of class $\mathcal{C}^1 \cap \text{Lip}$ on $\mathbb{D}$. For this step we shall use Lemma III.7.

Consider $u \in \mathbb{D}^m$, $v \in \mathbb{D}^n$, $F\colon \mathrm{R}^m \to \mathbb{R}$, $G\colon \mathbb{R}^n \to \mathbb{R}$, F, G of class $\mathcal{C}^1 \cap \mathrm{Lip}$.

According to this lemma, we can choose functions F_k, G_k of class $\mathcal{C}^\infty \cap \mathrm{Lip}$ such that

$$(*) \qquad \Gamma\big[F_k \circ u, G_k \circ v\big] = \sum_{ij} \frac{\partial F_k}{\partial x_i} \circ u \, \frac{\partial G_k}{\partial x_j} \circ v \, \Gamma\big[u_i, v_j\big]$$

and such that $F_k \to F$, $G_k \to G$ everywhere and $\frac{\partial F_k}{\partial x_i} \to \frac{\partial F}{\partial x_i}$ and $\frac{\partial G_k}{\partial x_j} \to \frac{\partial G}{\partial x_j}$ everywhere, with the functions $\frac{\partial F_k}{\partial x_i}$ and $\frac{\partial G_k}{\partial x_j}$ remaining bounded in modulus by a constant.

The right-hand side of $(*)$ converges to

$$\sum_{ij} \frac{\partial F}{\partial x_i} \circ u \cdot \frac{\partial G}{\partial x_j} \circ v \cdot \Gamma\big[u_i, v_j\big].$$

Moreover,

$$\mathcal{E}\big[F_k \circ u - F_{k'} \circ u\big] = \frac{1}{2} \int \sum_{ij} \Big(\frac{\partial F_k}{\partial x_i} \circ u - \frac{\partial F_{k'}}{\partial x_i} \circ u\Big)\Big(\frac{\partial F_k}{\partial x_j} \circ u - \frac{\partial F_{k'}}{\partial x_j} \circ u\Big) \Gamma\big[u_i, u_j\big]\, d\mathbb{P}$$

tends to zero by dominated convergence as $k, k' \uparrow \infty$.

By the closedness of $\mathcal{E}$, $F \circ u \in \mathbb{D}$ and $F_k \circ u \to F \circ u$ in $\mathbb{D}$ and, similarly, $G \circ v \in \mathbb{D}$ and $G_k \circ v \to G \circ v$ in $\mathbb{D}$. From the continuity of Γ, the left-hand side of $(*)$ tends to $\Gamma[F \circ u, G \circ v]$. This ends the proof of the functional calculus of class $\mathcal{C}^1 \cap \mathrm{Lip}$. ◇

Let us now present some closability results for specific error pre-structures.

First of all a complete answer can be provided for the closability question in dimension one. The following result is owed to M. Hamza (1975).

Definition III.20. Let $a\colon \mathbb{R} \to \mathbb{R}$ be a nonnegative measurable function. The set $R(a)$ of *regular points* of a is defined by

$$R(a) = \left\{x\colon \exists \varepsilon > 0 \int_{x-\varepsilon}^{x+\varepsilon} \frac{1}{a(t)}\, dt < +\infty\right\}.$$

In other words $R(a)$ is the largest open set V such that $\frac{1}{a} \in L^1_{\mathrm{loc}}(V, dt)$.

Proposition III.21. *Let m be a probability measure on $\mathbb{R}$. Let us set $\mathbb{D}_0 = \mathcal{C}_K^\infty(\mathbb{R})$ and for $u \in \mathbb{D}_0$*

$$\Gamma[u] = u'^2 \cdot g$$

where $g \geq 0$ is in $L^1_{\mathrm{loc}}(m)$. Then, the form $\mathcal{E}[u] = \frac{1}{2}\int u'^2 \cdot g \cdot dm$ is closable in $L^2(m)$ if and only if the measure $g \cdot m$ is absolutely continuous with respect to the Lebesgue measure dx and its density a vanishes dx-a.e. on $\mathbb{R} \setminus R(a)$.

For the proof we refer to Fukushima, Oshima, and Takeda [1994], p. 105.

Example III.22. Suppose m has density ρ with respect to dx, i.e. consider the pre-structure

$$\big(\mathbb{R}, \mathcal{B}(\mathcal{R}), \rho\, dx, \mathcal{C}_K^\infty(\mathbb{R}), \Gamma\big)$$

with $\Gamma[u] = u'^2 \cdot g$.

If the nonnegative functions ρ and g are *continuous*, then the pre-structure is closable.

Indeed, $a = \rho g$ is continuous and $\{a > 0\} \subset R(a)$, hence $a = 0$ on $\mathbb{R} \setminus R(a)$.

We now come to an important example, one of the historical applications of the theory of Dirichlet forms.

Example III.23. Let D be a connected open set in $\mathbb{R}^d$ with unit volume. Let $\mathbb{P} = dx$ be the Lebesgue measure on D. Let Γ be defined on $\mathcal{C}_K^\infty(D)$ via

$$\Gamma[u, v] = \sum_{ij} \frac{\partial u}{\partial x_i} \frac{\partial v}{\partial x_j} a_{ij}, \quad u, v \in \mathcal{C}_K^\infty(D),$$

where the functions a_{ij} are supposed to satisfy the following assumptions:

- $a_{ij} \in L^2_{\mathrm{loc}}(D) \quad \dfrac{\partial a_{ij}}{\partial x_k} \in L^2_{\mathrm{loc}}(D), \quad i, j, k = 1, \ldots, d,$
- $\sum_{i,j} a_{ij}(x)\xi_i\xi_j \geq 0 \quad \forall \xi \in \mathbb{R}^d \quad \forall x \in D,$
- $a_{ij}(x) = a_{ji}(x) \quad \forall x \in D.$

Then the pre-structure $\big(D, \mathcal{B}(D), \mathbb{P}, \mathcal{C}_K^\infty(D), \Gamma\big)$ is closable.

Proof. Consider the symmetric linear operator S from $\mathcal{C}_K^\infty(D)$ into $L^2(\mathbb{P})$ defined by

$$Su = \frac{1}{2} \sum_{i,j} \frac{\partial}{\partial x_i}\left(\frac{\partial u}{\partial x_j} a_{ij}\right), \quad u \in \mathcal{C}_K^\infty(D),$$

and the form $\mathcal{E}_0$ defined by

$$\mathcal{E}_0[u, v] = -\langle u, Sv\rangle_{L^2(\mathbb{P})} = -\langle Su, v\rangle_{L^2(\mathbb{P})}, \quad u, v \in \mathcal{C}_K^\infty(D).$$

The result is a consequence of the following lemma which is interesting in itself.

Lemma III.24. *Let $(\Omega, \mathcal{A}, \mathbb{P})$ be a probability space and $\mathbb{D}_0$ be a dense sub-vector space of $L^2(\mathbb{P})$. If S is a negative symmetric linear operator from $\mathbb{D}_0$ into $L^2(\mathbb{P})$, the positive symmetric bilinear form $\mathcal{E}_0$ defined by*

$$\mathcal{E}_0[u, v] = -\langle Su, v\rangle_{L^2(\mathbb{P})} = -\langle u, Sv\rangle_{L^2(\mathbb{P})}$$

for $u, v \in \mathbb{D}_0$ is closable.

Proof. Let $u_n \in \mathbb{D}_0$ be such that $\|u_n\|_{L^2(\mathbb{P})} \to 0$ as $n \uparrow \infty$, and $\mathcal{E}_0\big[u_n - u_m\big] \to 0$ as $n, m \uparrow \infty$.

Noting that for fixed m

$$\mathcal{E}_0\big[u_n, u_m\big] \xrightarrow[n\uparrow\infty]{} 0$$

the equality

$$\mathcal{E}_0\big[u_n\big] = \mathcal{E}_0\big[u_n - u_m\big] + 2\mathcal{E}_0\big[u_n, u_m\big] - \mathcal{E}_0\big[u_m\big]$$

shows that the real number $\mathcal{E}_0\big[u_n\big]$ converges to a limit which is necessarily zero. This proves that $\mathcal{E}_0$ is closable. ◇

This result paves the way for constructing a semigroup and a Markov process with generator S, without any assumption of regularity of the matrix $\big(a_{ij}(x)\big)_{ij}$. (See, for example Ma–Röckner (1992).)

Example III.25 (Classical case with minimal domain). Let D be a connected open set in $\mathbb{R}^d$ with unit volume. Applying the preceding result to the case $a_{ij} = \delta_{ij}$, we find $S = \frac{1}{2}\Delta$ the Laplacian operator and $\Gamma[u] = |\nabla u|^2$. The completion of $\mathcal{C}_K^\infty(D)$ for the norm $\|u\|_{\mathbb{D}} = \big(\|u\|_{L^2}^2 + \|\nabla u\|_{L^2}^2\big)^{1/2}$ is the space usually denoted $H_0^1(D)$. We obtain the same error structure $\big(D, \mathcal{B}(D), dx, H_0^1(D), |\nabla \cdot|^2\big)$ as in Example 1.3 but with a smaller domain for Γ: H_0^1 instead of H^1. ◇

Let us now give two examples of non-closable pre-structures.

Example III.26. Consider the following

$$\Omega = [0, 1]$$
$$\mathcal{A} = \mathcal{B}\big(]0, 1[\big)$$
$$\mathbb{P} = \frac{1}{2}\delta_0 + \sum_{n=1}^{\infty} \frac{1}{2^{n+1}}\delta_{\frac{1}{n}}$$

where δ_a denotes the Dirac mass at a. Let us choose for $\mathbb{D}$ the space

$$\mathbb{D} = \Big\{f : [0, 1] \to \mathbb{R} \text{ such that } f \in L^2(\mathbb{P}) \text{ and } f'(0) = \lim_{n\to\infty} \frac{f\big(\frac{1}{n}\big) - f(0)}{\frac{1}{n}} \text{ exists}\Big\}$$

and for Γ

$$\Gamma[f] = \big(f'(0)\big)^2 \quad \text{for } f \in \mathbb{D}.$$

The term $(\Omega, \mathcal{A}, \mathbb{P}, \mathbb{D}, \Gamma)$ is an error pre-structure. But the form $\mathcal{E}[f] = \frac{1}{2}\mathbb{E}\big[\Gamma[f]\big]$ is not closable, since it is easy to find a sequence $f_n \in \mathbb{D}$ such that $f_n'(0) = 1$ and $f_n \to 0$ in $L^2(\mathbb{P})$. Such a sequence prevents the closability condition (Definition III.17) from being satisfied.

Example III.27. Let Ω be $\big(]0, 1[\big)^2$ equipped with its Borelian subsets. Let us take for $\mathbb{P}$ the following probability measure

$$\mathbb{P} = \frac{1}{2}\, dx_1\, dx_2 + \frac{1}{2}\mu$$

where μ is the uniform probability on the diagonal of the square. Define Γ on $\mathcal{C}_K^1(\Omega)$ by

$$\Gamma[f] = \left(\frac{\partial f}{\partial x_1}\right)^2 + \left(\frac{\partial f}{\partial x_2}\right)^2.$$

Then the error pre-structure

$$\big(]0, 1]^2, \mathcal{B}(]0, 1[^2), \mathbb{P}, \mathcal{C}_K^1(]0, 1[^2), \Gamma\big)$$

is not closable.

Indeed, it is possible (exercise) to construct a sequence $f_n \in \mathcal{C}_K^\infty(\Omega)$ such that

$$(*) \qquad -\frac{1}{n} \le f_n \le \frac{1}{n}$$

$$(**) \qquad f_n(x_1, x_2) = x_2 - x_1$$

on a neighborhood of the diagonal of $\big([\frac{1}{n}, 1 - \frac{1}{n}[\big)^2$. By $(*)$ the sequence f_n tends to zero in $L^2(\mathbb{P})$ and by $(**)$

$$\left|\nabla f_n\right|^2 \to 2 \quad \mu\text{-a.e.}$$

Hence

$$\underline{\lim}_n\, \mathcal{E}\big[f_n\big] = \frac{1}{2}\, \underline{\lim}_n\, \mathbb{E}\left|\nabla f_n\right|^2 \ge 1.$$

This statement contradicts the closability condition. ◇

For an error pre-structure, closability is a specific link between the operator Γ and the measure $\mathbb{P}$. Only sufficient conditions for closability are known in dimension greater or equal to 2, and any program of improving these conditions toward a necessary and sufficient characterization is tricky. Fortunately, closedness is preserved by the two important operations of taking products and taking images, as we shall see in the following chapter. This feature provides a large number of closed structures for stochastic analysis.

Bibliography for Chapter III

N. Bouleau, Décomposition de l'énergie par niveau de potentiel, in *Théorie du Potentiel, Orsay 1983,* Lecture Notes in Math. 1096, Springer-Verlag, 1984, 149–172.

N. Bouleau and F. Hirsch, *Dirichlet Forms and Analysis on the Wiener Space*, Walter de Gruyter, 1991.

M. Fukushima, Y. Oshima and M. Takeda, *Dirichlet Forms and Markov Processes*, Walter de Gruyter, 1994.

Z.-M. Ma and M. Röckner, *Introduction to the Theory of (Non-symmetric) Dirichlet Forms*, Springer-Verlag, 1992.

P. Malliavin, *Stochastic Analysis*, Springer-Verlag, 1997.

P. Mattila, *Geometry of Sets and Measures in Euclidean Spaces*, Cambridge University Press, 1995.

F. Morgan, *Geometric Measure Theory, A Beginner's Guide*, Academic Press, 1988.

N. Nualart, *The Malliavin Calculus and Related Topics*, Springer-Verlag, 1995.

W. Rudin, *Real and Complex Analysis*, McGraw-Hill, 1966.

W. Rudin, *Functional Analysis*, McGraw-Hill, 1973.

L. Schwartz, *Théorie des distributions*, Hermann, 1966.

Chapter IV

Images and products of error structures

In Chapter I, we provided an application with a Markov chain represented as

$$X_{n+1}^x = \Psi\big(X_n^x, U_{n+1}\big), \quad X_0^x = x,$$

where $U_1, \dots, U_n, \dots$ was a sequence of i.i.d. random variables uniformly distributed on $[0, 1]$ representing the Monte Carlo samples. Several interesting questions deal with all of the X_n's, e.g. the limit $\lim_N \frac{1}{N} \sum_{n=1}^{N} f\big(X_n^x\big)$. Under such circumstances, if we consider the U_n's to be erroneous, we must construct an error structure on the infinite product $[0, 1]^{\mathbb{N}}$ equipped with the product probability measure. Once this has been accomplished, $X_1^x = \Psi\big(x, U_1\big)$, $X_2^x = \Psi\big(X_1^x, U_2\big), \dots$ will also be erroneous as an image by Ψ of erroneous quantities.

This requires two operations that will be studied in this chapter: defining both the product of error structures and the image of an error structure by an application.

1 Images

Let $S = (\Omega, \mathcal{A}, \mathbb{P}, \mathbb{D}, \Gamma)$ be an error structure; consider an $\mathbb{R}^d$-valued random variable $X \colon \Omega \to \mathbb{R}^d$ such that $X \in \mathbb{D}^d$, i.e. $X = \big(X_1, \dots, X_d\big)$, $X_i \in \mathbb{D}$ for $i = 1, \dots, d$. We will define the *error structure image of* S *by* X.

First of all, the probability space on which this error structure will be defined is the image of $(\Omega, \mathcal{A}, \mathbb{P})$ by X:

$$(\Omega, \mathcal{A}, \mathbb{P}) \xrightarrow{X} \big(\mathbb{R}^d, \mathcal{B}(\mathbb{R}^d), X_*\mathbb{P}\big),$$

where $X_*\mathbb{P}$ is the law of X, i.e. the measure such that $(X_*\mathbb{P})(E) = \mathbb{P}\big(X^{-1}(E)\big)$ $\forall E \in \mathcal{B}(\mathbb{R}^d)$.

We may then set

$$\mathbb{D}_X = \big\{u \in L^2(X_*\mathbb{P}) \colon u \circ X \in \mathbb{D}\big\}$$

and

$$\Gamma_X[u](x) = \mathbb{E}\big[\Gamma[u \circ X] \mid X = x\big].$$

(Unless explicitly mentioned otherwise, the symbol $\mathbb{E}$ denotes the expectation or conditional expectation with respect to $\mathbb{P}$.)

Comment. Let us recall that if Y is an integrable random variable, the random variable $\mathbb{E}[Y \mid X]$ (which is, by definition, $\mathbb{E}[Y \mid \sigma(X)]$ where $\sigma(X)$ is the σ-field generated by X) is a function of X: a Borel function φ exists such that

$$\mathbb{E}[Y \mid X] = \varphi(X),$$

and the notation $\mathbb{E}[Y \mid X = x]$ means $\varphi(x)$. As easily shown, the function φ is unique up to an a.e. equality with respect to the law of X.

If we denote the image of a probability measure μ by a random variable X as $X_*\mu$, the function φ can be defined as the density of the measure $X_*(Y \cdot \mathbb{P})$ with respect to the measure $X_*\mathbb{P}$:

$$\varphi = \frac{dX_*(Y \cdot \mathbb{P})}{dX_*\mathbb{P}}.$$

This expression is a direct consequence of the definition of the conditional expectation.

The definition of Γ can thus be written

$$\Gamma_X[u] = \frac{dX_*\big(\Gamma[u \circ X] \cdot \mathbb{P}\big)}{dX_*\mathbb{P}}$$

and we also remark that

$$\Gamma_X[u](X) = \mathbb{E}\big[\Gamma[u \circ X] \mid X\big].$$

Proposition IV.1. $X_*S = \big(\mathbb{R}^d, \mathcal{B}(\mathbb{R}^d), X_*\mathbb{P}, \mathbb{D}_X, \Gamma_X\big)$ *is an error structure, the coordinate maps of* $\mathbb{R}^d$ *are in* $\mathbb{D}_X$, *and* X_*S *is Markovian if* S *is Markovian.*

Proof. 1) $\mathbb{D}_X$ contains Lipschitz functions and hence is dense in $L^2(X_*\mathbb{P})$.

2) Let us begin by proving the closedness before the functional calculus: Is the form $\mathcal{E}_X[u] = \frac{1}{2}\int \Gamma_X[u]\, d(X_*\mathbb{P})$ closed?

Let us remark that

$$\mathcal{E}_X[u] = \frac{1}{2}\mathbb{E}\big[\Gamma[u \circ X]\big] = \mathcal{E}[u \circ X].$$

Then, if u_n is a Cauchy sequence in $\mathbb{D}_X$

a) $\exists u$ such that $u_n \to u$ in $L^2(X_*\mathbb{P})$

b)

$$\begin{aligned} \|u_n - u_m\|_{\mathbb{D}_X} &= \big(\|u_n - u_m\|^2_{L^2(X_*\mathbb{P})} + \mathcal{E}_X[u_n - u_m]\big)^{1/2} \\ &= \big(\|u_n \circ X - u_m \circ X\|^2_{L^2(\mathbb{P})} + \mathcal{E}[u_n \circ X - u_m \circ X]\big)^{1/2} \\ &= \|u_n \circ X - u_m \circ X\|_{\mathbb{D}} \end{aligned}$$

hence $u_n \circ X$ is Cauchy in $\mathbb{D}$ and by a), $u_n \circ X \to u \circ X$ in $L^2(\mathbb{P})$, which implies by the closedness of $\mathcal{E}$ that $u \circ X \in \mathbb{D}$ and $u_n \circ X \to u \circ X$ in $\mathbb{D}$. In other words, $u \in \mathbb{D}_X$ and $u_n \to u$ in $\mathbb{D}_X$.

3) In order to prove the functional calculus of class $\mathcal{C}^1 \cap \text{Lip}$, let as usual $u \in \mathbb{D}_X^m$, $v \in \mathbb{D}_X^n$ and F, G be of class $\mathcal{C}^1 \cap \text{Lip}$. We then obtain the $\mathbb{P}$-a.e. equalities

$$\begin{aligned}
\Gamma_X[F \circ u, G \circ v](X) &= \mathbb{E}\big[\Gamma[F \circ u \circ X, G \circ v \circ X] \mid X\big] \\
&= \mathbb{E}\Big[\sum_{i,j} \frac{\partial F}{\partial x_i} \circ u \circ X \frac{\partial G}{\partial x_j} \circ v \circ X \Gamma\big[u_i \circ X, v_j \circ X\big] \mid X\Big] \\
&= \sum_{i,j} \frac{\partial F}{\partial x_i} \circ u \circ X \frac{\partial G}{\partial x_j} \circ v \circ X\, \mathbb{E}\big[\Gamma[u_i \circ X, v_j \circ X] \mid X\big] \\
&= \sum_{i,j} \frac{\partial F}{\partial x_i} \circ u \circ X \frac{\partial G}{\partial x_j} \circ v \circ X\, \Gamma_X\big[u_i, v_j\big](X).
\end{aligned}$$

This can be also written as follows:

$$\Gamma_X[F \circ u, G \circ v](x) = \sum_{i,j} \frac{\partial F}{\partial x_i} \circ u(x) \frac{\partial G}{\partial x_j} \circ v(x) \Gamma_X\big[u_i, v_j\big](x) \text{ for } (X_*\mathbb{P})\text{-a.e. } x. \quad \diamond$$

Example IV.2. Consider the open sector $\Omega = \left(\mathbb{R}_+^*\right)^2$ with the Borel σ-field and the probability measure

$$\mathbb{P}(dx, dy) = e^{-x-y}\, dx\, dy.$$

On the domain $\mathbb{D}_0 = \mathcal{C}_b^1(\Omega)$ (bounded $\mathcal{C}^1$-functions with bounded derivatives) we consider the operator

$$\Gamma[u] = \left(\left(\frac{\partial u}{\partial x}\right)^2 + \left(\frac{\partial u}{\partial y}\right)^2\right) \cdot g,$$

g being bounded continuous and strictly positive. The pre-structure $\left(\Omega, \mathcal{A}, \mathbb{P}, \mathbb{D}_0, \Gamma\right)$ is shown to be closable, let $S = (\Omega, \mathcal{A}, \mathbb{P}, \mathbb{D}, \Gamma)$ be the associate error structure.

What is the image U_*S of this structure by the application $U(x, y) = (x \wedge y, x \vee y - x \wedge y)$? (Here $x \wedge y = \min(x, y)$ and $x \vee y = \max(x, y)$.)

a) Let us first check $U \in \mathbb{D}$. Let us denote the coordinate maps by capital letters $X(x, y) = x$, $Y(x, y) = y$. With this notation,

$$U = (X \wedge Y, X \vee Y - X \wedge Y).$$

As in Property 2.2 of Chapter III, it is easily demonstrated that $n \operatorname{Arctan} \frac{X}{n} \in \mathbb{D}_0$ and is Cauchy for the $\mathbb{D}$-norm, hence $X \in \mathbb{D}$. Similarly, $Y \in \mathbb{D}$ and U, as a Lipschitz function of elements of $\mathbb{D}$, belongs to $\mathbb{D}$.

b) It becomes an elementary exercise in probability calculus to prove that the law of U is the measure

$$U_*\mathbb{P} = 2e^{-2s}\, ds \cdot e^{-t}\, dt \quad \text{on } \left(\mathbb{R}_+^*\right)^2$$

and that $X \wedge Y$ and $X \vee Y - X \wedge Y$ are independent random variables.

c) Computing $\Gamma[X \wedge Y]$ is performed thanks to functional calculus (Proposition III.15 of Chapter III) by using the fact that the law of (X, Y) has a density:

$$\Gamma[X \wedge Y] = 1_{\{X<Y\}} \cdot g + 1_{\{X>Y\}} \cdot g = g.$$

Similarly,

$$\Gamma[X \vee Y] = g$$

and

$$\Gamma[X \vee Y, X \wedge Y] = 0.$$

The matrix

$$\underline{\underline{\Gamma}}[U] = \begin{pmatrix} \Gamma[U_1] & \Gamma[U_1, U_2] \\ \Gamma[U_1, U_2] & \Gamma[U_2] \end{pmatrix}$$

is given by

$$\underline{\underline{\Gamma}}[U] = gA,$$

where $A = \begin{pmatrix} 1 & -1 \\ -1 & 2 \end{pmatrix}$. If $F \in \mathcal{C}^1 \cap \mathrm{Lip}$,

$$\Gamma[F \circ U] = (\nabla F)^t \circ U \cdot A \cdot \nabla F \circ U \cdot g$$

and

$$\mathbb{E}\big[\Gamma[F \circ U] \mid U = (s,t)\big] = (\nabla F)^t(s,t) \cdot A \cdot \nabla F(s,t)\mathbb{E}[g \mid U = (s,t)].$$

Computing $\mathbb{E}[g \mid U = (s,t)]$ is a purely probabilistic exercise and yields

$$\mathbb{E}[g \mid U = (s,t)] = \frac{g(s, s+t) + g(s+t, s)}{2}.$$

Finally, the image error structure is written as follows:

$$\big(\Omega, \mathcal{A}, 2e^{-2s} e^{-t}\, ds\, dt, \mathbb{D}_U, \Gamma_U\big)$$

with

$$\Gamma_U[F](s,t) = \left[\left(\frac{\partial F}{\partial s}(s,t)\right)^2 - 2\frac{\partial F}{\partial s}\frac{\partial F}{\partial t}(s,t) + 2\left(\frac{\partial F}{\partial t}(s,t)\right)^2\right] \cdot \frac{g(s, s+t) + g(s+t, s)}{2}.$$

If X and Y are machine breakdown times, the hypotheses indicate that the pairs $(X, \text{error on } X)$ and $(Y, \text{error on } Y)$ are independent (see Chapter I). Thus $U_1 = X \wedge Y$ is the time of the first breakdown and $U_2 = X \vee Y - X \wedge Y$ the time to wait between the first and second breakdowns. It is well-known that U_1 and U_2 are independent, but we recognize that the error on U_1 and the error on U_2 are linked. ◇

Remark. If in our construction of the image structure X_*S the random variable X is no longer supposed to be in $\mathbb{D}$, but simply to be measurable, the entire argument still holds, except for the density of $\mathbb{D}_X$. We thus obtain:

Proposition IV.3. *Let X be a measurable map from $(\Omega, \mathcal{A})$ into a measurable space $(E, \mathcal{F})$. If $\mathbb{D}_X$ is dense in $L^2(E, \mathcal{F}, X_*\mathbb{P})$, then*

$$\big(E, \mathcal{F}, X_*\mathbb{P}, \mathbb{D}_X, \Gamma_X\big)$$

defined as before is an error structure.

Example IV.4. Consider the Cauchy law on $\mathbb{R}$

$$\mathbb{P} = \frac{a\,dx}{\pi(a^2 + x^2)}$$

and the error structure

$$S = \big(\mathbb{R}, \mathcal{B}(\mathbb{R}), \mathbb{P}, \mathbb{D}, \Gamma\big)$$

where $\Gamma[u](x) = u'^2(x)\alpha^2(x)$ for $u \in \mathcal{C}_b^1(\mathbb{R})$ (space of bounded functions of class $\mathcal{C}^1$ with bounded derivative) and $(\mathbb{D}, \Gamma)$ is the smallest closed extension of $\big(\mathcal{C}_b^1(\mathbb{R}), \Gamma\big)$. The function α is assumed continuous and bounded.

We want to study the image of S by the mapping $U : x \to \{x\}$, where $\{x\}$ denotes the fractional part of x. Clearly, U does not belong to $\mathbb{D}$ since U is discontinuous at integer points and functions in $\mathbb{D}$ can be shown as continuous (as soon as α does not vanish).

a) To compute the image of $\mathbb{P}$ by U, let us take a Borel bounded function f. We have

$$\mathbb{E}\big[f(\{x\})\big] = \int_0^1 f(t) \sum_{n\in\mathbb{Z}} \frac{a\,dt}{\pi\big(a^2 + (t+n)^2\big)}.$$

According to the Poisson sum formula (see L. Schwartz *Théorie des distributions*, p. 254):

$$\begin{aligned}
\sum_{n\in\mathbb{Z}} \frac{a}{\pi\big(a^2 + (t+n)^2\big)} &= \sum_{n\in\mathbb{Z}} e^{2i\pi nt} e^{-2\pi|n|a} \\
&= \frac{1}{1 - e^{2\pi(it-a)}} + \frac{1}{1 - e^{2\pi(-it-a)}} - 1 \\
&= \frac{\sinh 2\pi a}{\cosh 2\pi a - \cos 2\pi t}.
\end{aligned}$$

It follows that the image measure $U_*\mathbb{P}$ is the probability measure

$$\frac{\sinh 2\pi a}{\cosh 2\pi a - \cos 2\pi t}\,dt \quad \text{on } [0,1].$$

b) The domain

$$\mathbb{D}_U = \{g \in L^2(U_*\mathbb{P}) : g \circ U \in \mathbb{D}\}$$

contains the function g of class $\mathcal{C}^1$ on $[0, 1]$ such that $g(0) = g(1)$ and $g'(0) = g'(1)$. It is therefore dense in $L^2(U_*\mathbb{P})$.

c) In order to compute Γ_U, let us consider a function g as above and then evaluate $\mathbb{E}[\Gamma[g \circ U] \mid U]$.

Coming back to the definition of the conditional expectation, we must calculate the following for a Borel bounded function h:

$$\mathbb{E}[\Gamma[g \circ U]h(U)] = \sum_{n\in\mathbb{Z}} \int_0^1 g'^2(t)\alpha^2(t+n)h(t)\frac{a\,dt}{\pi\left(a^2 + (t+n)^2\right)}.$$

Writing this expression as follows

$$= \int_0^1 g'^2(t)h(t)\frac{\sum_n \frac{a\alpha^2(t+n)}{\pi\left(a^2+(t+n)^2\right)}}{\sum_n \frac{a}{\pi\left(a^2+(t+n)^2\right)}} \sum_n \frac{a}{\pi\left(a^2+(t+n)^2\right)}\,dt$$

yields

$$\Gamma_U[g](t) = g'^2(t)\frac{\sum_n \frac{a\alpha^2(t+n)}{\pi\left(a^2+(t+n)^2\right)}}{\sum_n \frac{a}{\pi\left(a^2+(t+n)^2\right)}}.$$

In the case $\alpha^2(x) = \frac{a^2+x^2}{b^2+x^2}$ for example, we obtain

$$\Gamma_U[g] = g'^2(t)\frac{a}{b}\,\frac{\cosh 2\pi a - \cos 2\pi t}{\cosh 2\pi b - \cos 2\pi t} \cdot \frac{\sinh 2\pi b}{\sinh 2\pi a}. \qquad \diamond$$

2 Finite products

Let $S_1 = (\Omega_1, \mathcal{A}_1, \mathbb{P}_1, \mathbb{D}_1, \Gamma_1)$ and $S_2 = (\Omega_2, \mathcal{A}_2, \mathbb{P}_2, \mathbb{D}_2, \Gamma_2)$ be two error structures.

The aim then is to define on the product probability space

$$(\Omega, \mathcal{A}, \mathbb{P}) = (\Omega_1 \times \Omega_2, \mathcal{A}_1 \otimes \mathcal{A}_2, \mathbb{P}_1 \times \mathbb{P}_2)$$

an operator Γ and its domain $\mathbb{D}$ in such a way that $(\Omega, \mathcal{A}, \mathbb{P}, \mathbb{D}, \Gamma)$ is an error structure expressing the condition that the two coordinate mappings and their errors are independent (see Chapter 1, Section 2.2).

Proposition IV.5. *Let us define*

$$(\Omega, \mathcal{A}, \mathbb{P}) = \big(\Omega_1 \times \Omega_2, \mathcal{A}_1 \otimes \mathcal{A}_2, \mathbb{P}_1 \times \mathbb{P}_2\big),$$

$$\mathbb{D} = \Big\{ f \in L^2(\mathbb{P}) : \text{ for } \mathbb{P}_1\text{-a.e. } x\ f(x,\cdot) \in \mathbb{D}_2 \text{for } \mathbb{P}_2\text{-a.e. } y\ f(\cdot, y) \in \mathbb{D}_1$$
$$\text{and } \int \big(\Gamma_1[f(\cdot, y)](x) + \Gamma_2\big[f(x,\cdot)\big](y) d\mathbb{P}_1(x)\, d\mathbb{P}_2(y) < +\infty \Big\},$$

and for $f \in \mathbb{D}$

$$\Gamma[f](x, y) = \Gamma_1[f(\cdot, y)](x) + \Gamma_2[f(x,\cdot)](y).$$

Then $S = (\Omega, \mathcal{A}, \mathbb{P}, \mathbb{D}, \Gamma)$ *is an error structure denoted* $S = S_1 \times S_2$ *and called the product of* S_1 *and* S_2*, whereby* S *is Markovian if* S_1 *and* S_2 *are both Markovian.*

Proof. 1) From the construction of the product measure, we know that functions of the form

$$\sum_{i=1}^{n} u_i(x) v_i(y) \quad u_i \in L^2(\mathbb{P}_1) \quad v_i \in L^2(\mathbb{P}_2)$$

are dense in $L^2(\mathbb{P})$. They can be approximated in $L^2(\mathbb{P})$ by functions of the form

$$\sum_{i=1}^{n} \alpha_i(x) \beta_i(y), \quad \alpha_i \in \mathbb{D}_1,\ \beta_i \in \mathbb{D}_2,$$

which are in $\mathbb{D}$. Hence, $\mathbb{D}$ is dense in $L^2(\mathbb{P})$.

2) The functional calculus of class $\mathcal{C}^1 \cap \mathrm{Lip}$ for Γ is straightforward from the definition.

3) Is the form $\mathcal{E}[f] = \int \big(\mathcal{E}_1[f] + \mathcal{E}_2[f]\big)\, d\mathbb{P}$ associated with Γ closed?

To see this, let (f_n) be a Cauchy sequence in $\mathbb{D}$ equipped, as usual, by the norm:

$$\|\cdot\|_{\mathbb{D}} = \big(\|\cdot\|_{L^2}^2 + \mathcal{E}[\cdot]\big)^{1/2}.$$

There exists an $f \in L^2(\mathbb{P})$ such that $f_n \to f$ in $L^2(\mathbb{P})$, and there exists a subsequence f_{n_k} such that

$$\begin{cases} \sum_k \|f_{n_k} - f\|_{L^2}^2 < +\infty \\ \sum_k \big(\mathcal{E}\left[f_{n_{k+1}} - f_{n_k}\right]\big)^{1/2} < +\infty. \end{cases}$$

It follows that for $\mathbb{P}_1$-a.e. x, we have

$$\int \sum_k \left|f_{n_k}(x, y) - f(x, y)\right|^2 d\mathbb{P}_2(y) < +\infty$$

and

$$\sum_k \left(\mathcal{E}_2\left[f_{n_{k+1}}(x,\cdot) - f_{n_k}(x,\cdot)\right]\right)^{1/2} < +\infty.$$

(The second inequality stems from the remark that if

$$\mathcal{E}[g_k] = \mathbb{E}_1\mathcal{E}_2[g_k] + \mathbb{E}_2\mathcal{E}_1[g_k]$$

then

$$\left(\mathcal{E}[g_k]\right)^{1/2} \geq \left(\mathbb{E}_1\mathcal{E}_2[g_k]\right)^{1/2} \geq \mathbb{E}_1\left(\mathcal{E}_2[g_k]\right)^{1/2}.)$$

this implies that $f_{n_k}(x,\cdot)$ is Cauchy in $\mathbb{D}_2$. (Indeed, the condition $\sum_k \|a_{k+1} - a_k\| < +\infty$ implies that the sequence (a_k) is Cauchy.)

Since the form $\mathcal{E}_2$ is closed in $L^2(\mathbb{P}_2)$, we obtain that for $\mathbb{P}_1$-a.e. x

$$f(x,\cdot) \in \mathbb{D}_2 \text{ and } f_{n_k}(x,\cdot) \to f(x,\cdot) \text{ in } \mathbb{D}_2.$$

Similarly, for $\mathbb{P}_2$-a.e. y

$$f(\cdot,y) \in \mathbb{D}_1 \text{ and } f_{n_k}(\cdot,y) \to f(\cdot,y) \text{ in } \mathbb{D}_1.$$

With this preparation now being complete, we will see that the main argument of the proof is provided by the Fatou lemma in integration theory:

$$\begin{aligned}
&\frac{1}{2}\int \left(\Gamma_1[f] + \Gamma_2[f]\right) d\mathbb{P}_1\, d\mathbb{P}_2 \\
&\quad = \int \left(\mathcal{E}_1[f]\right)(y)\, d\mathbb{P}_2(y) + \int \left(\mathcal{E}_2[f]\right)(x)\, d\mathbb{P}_1(x) \\
&\quad = \int \lim_k \mathcal{E}_1\left[f_{n_k}\right](y)\, d\mathbb{P}_2(y) + \int \lim_k \mathcal{E}_2\left[f_{n_k}\right](x)\, d\mathbb{P}_1(x).
\end{aligned}$$

According to the Fatou lemma, we can put the limits outside the integrals as liminf

$$\leq 2 \varliminf_k \ \mathcal{E}\left[f_{n_k}\right];$$

this is $< +\infty$ since f_n is Cauchy in $\mathbb{D}$ and a Cauchy sequence is always bounded. Thus, $f \in \mathbb{D}$.

We can then write the following:

$$\begin{aligned}
\mathcal{E}\left[f_n - f\right] &= \int \mathcal{E}_1\left[f_n - f\right] d\mathbb{P}_2(y) + \int \mathcal{E}_2\left[f_n - f\right] d\mathbb{P}_1(x) \\
&= \int \lim_k \mathcal{E}_1\left[f_n - f_{n_k}\right](y)\, d\mathbb{P}_2(y) + \int \lim_k \mathcal{E}_2\left[f_n - f_{n_k}\right](x)\, d\mathbb{P}_2 \\
&\leq 2 \varliminf_k \ \mathcal{E}\left[f_n - f_{n_k}\right]
\end{aligned}$$

using, once again, the Fatou lemma. Yet, this can be made as small as we want for large n since (f_n) is $\mathbb{D}$-Cauchy. This proves that $\mathcal{E}$ is closed.

The proof of the proposition has been accomplished. $\diamond$

The case of finite products is obtained similarly. We write the statement for the notation.

Proposition IV.6. *Let $S_n = (\Omega_n, \mathcal{A}_n, \mathbb{P}_n, \mathbb{D}_n, \Gamma_n)$ be error structures. The finite product*

$$S = \left(\Omega^{(N)}, \mathcal{A}^{(N)}, \mathbb{P}^{(N)}, \mathbb{D}^{(N)}, \Gamma^{(N)}\right) = \prod_{n=1}^{N} S_n$$

is defined as follows:

$$\begin{aligned}
\left(\Omega^{(N)}, \mathcal{A}^{(N)}, \mathbb{P}^{(N)}\right) &= \Big(\prod_{n=1}^{N} \Omega_n, \otimes_{n=1}^{N} \mathcal{A}_n, \prod_{n=1}^{N} \mathbb{P}_n\Big) \\
\mathbb{D}^{(N)} &= \Big\{ f \in L^2\left(\mathbb{P}^{(N)}\right) : \forall n = 1, \dots, N \\
&\qquad \textit{for } \mathbb{P}_1 \times \mathbb{P}_2 \times \dots \times \mathbb{P}_{n-1} \times \mathbb{P}_{n+1} \times \dots \times \mathbb{P}_N\text{-a.e.} \\
&\qquad w_1, w_2, \dots, w_{n-1}, w_{n+1}, \dots, w_N \text{ the function} \\
&\qquad x \to f\left(w_1, w_2, \dots, w_{n-1}, x, w_{n+1}, \dots, w_N\right) \in \mathbb{D}_n \\
&\qquad \text{and } \int \left(\Gamma_1[f] + \Gamma_2[f] + \dots + \Gamma_N[f]\right) d\mathbb{P} < +\infty \Big\}
\end{aligned}$$

and for $f \in \mathbb{D}$

$$\Gamma[f] = \Gamma_1[f] + \dots + \Gamma_N[f]$$

(where Γ_i applied to f is assumed to act only on the i-th variable of f.) S is an error structure, Markovian if the S_n's are Markovian.

We can now study the case of infinite products.

3 Infinite products

We will begin with a lemma showing that the limits of error structures on the same space with increasing quadratic error operators give rise to error structures.

Lemma IV.7. *Let $(\Omega, \mathcal{A}, \mathbb{P}, \mathbb{D}_i, \Gamma_i)$ be error structures, $i \in \mathbb{N}$, such that for $i < j$*

$$\mathbb{D}_i \supset \mathbb{D}_j \quad \textit{and} \quad \forall f \in \mathbb{D}_j \;\; \Gamma_i[f] \le \Gamma_j[f].$$

Let

$$\mathbb{D} = \Big\{ f \in \bigcap_i \mathbb{D}_i : \lim_i \uparrow \mathbb{E}\Gamma_i[f] < +\infty \Big\}$$

and for $f \in \mathbb{D}$, let $\Gamma[f] = \lim_i \uparrow \Gamma_i[f]$. Then $(\Omega, \mathcal{A}, \mathbb{P}, \mathbb{D}, \Gamma)$ is an error structure as soon as $\mathbb{D}$ is dense in $L^2(\mathbb{P})$.

Proof. Let us first remark of all that if $f \in \mathbb{D}$, then $\Gamma_i[f] \to \Gamma[f]$ in $L^1(\mathbb{P})$, since

$$\Gamma[f] = \lim_i \uparrow \Gamma_i[f] \quad \text{and} \quad \lim_i \mathbb{E}\Gamma_i[f] < +\infty.$$

a) Let us begin with the closedness of the form $\mathcal{E}[f] = \frac{1}{2}\mathbb{E}\Gamma[f]$. Let (f_n) be a Cauchy sequence in $\mathbb{D}$ (with $\|\cdot\|_{\mathbb{D}}$), then (f_n) is Cauchy in $\mathbb{D}_i$ (with $\|\cdot\|_{\mathbb{D}_i}$). If f is the limit of f_n in $L^2(\mathbb{P})$, we observe that $f_n \to f$ in $\mathbb{D}_i$ *uniformly with respect to i* and this implies $f_n \to f$ in $\mathbb{D}$.

Let us explain the argument further. We start from

$$\forall \varepsilon > 0\ \exists N, \quad p, q \geq N \Rightarrow \|f_p - f_q\|_{\mathbb{D}} \leq \varepsilon.$$

We easily deduce that $f_k \xrightarrow[k\uparrow\infty]{} f$ in $\mathbb{D}_i$. Let us now consider $n \geq N$. In the inequality

$$\|f_n - f\|_{\mathbb{D}_i} \leq \|f_n - f_q\|_{\mathbb{D}_i} + \|f_q - f\|_{\mathbb{D}_i}$$

we are free to choose q as we want, hence

$$\|f_n - f\|_{\mathbb{D}_i} \leq \varepsilon \quad \forall i$$

and

$$\frac{1}{2}\mathbb{E}\Gamma\big[f_n - f\big] = \lim_i \uparrow \mathbb{E}\Gamma_i\big[f_n - f\big] \leq \lim_i \uparrow \|f_n - f\|^2_{\mathbb{D}_i} \leq \varepsilon.$$

This proves that $f_n \to f$ in $\mathbb{D}$.

b) In order to prove the $\mathcal{C}^1 \cap \mathrm{Lip}$-functional calculus, let us consider as usual

$$u \in \mathbb{D}^m, \quad v \in \mathbb{D}^n, \quad F, G \in \mathcal{C}^1 \cap \mathrm{Lip}.$$

We know that $F \circ u \in \mathbb{D}_k$, $G \circ v \in \mathbb{D}_k$, and

$$\Gamma_k[F \circ u] = \sum_{ij} F_i' \circ u\, F_j' \circ v\, \Gamma_k\big[u_i, v_j\big].$$

From

$$\Gamma_k\big[u_i, v_j\big] = \frac{1}{2}\big[\Gamma_k\big[u_i + v_j\big] - \Gamma_k\big[u_i\big] - \Gamma_k\big[v_j\big]\big]$$

we see that $\Gamma_k\big[u_i, v_j\big] \to \Gamma\big[u_i, v_j\big]$ in $L^1(\mathbb{P})$. Hence

$$\lim_k \uparrow \Gamma_k[F \circ u] = \sum_{ij} F_i' \circ u\, F_j' \circ u\, \Gamma\big[u_i, u_j\big] \in L^1(\mathbb{P})$$

which implies that $F \circ u \in \mathbb{D}$ and

$$\Gamma[F \circ u] = \sum_{ij} F_i' \circ u\, F_j' \circ v\, \Gamma\big[u_i, v_j\big].$$

The equality

$$\Gamma[F \circ u, G \circ v] = \sum_{ij} F_i' \circ u\, G_j' \circ v\, \Gamma\big[u_i, v_j\big]$$

then follows by means of polarization. ◇

Let us apply this lemma to the case of infinite products. Let $S_n = \big(\Omega_n, \mathcal{A}_n, \mathbb{P}_n, \mathbb{D}_n, \Gamma_n\big)$ be error structures. The finite products

$$\prod_{n=1}^{N} S_n = \big(\Omega^{(N)}, \mathcal{A}^{(N)}, \mathbb{P}^{(N)}, \mathbb{D}^{(N)}, \Gamma^{(N)}\big)$$

have already been defined. On

$$(\Omega, \mathcal{A}, \mathbb{P}) = \Big(\prod_{n=1}^{\infty} \Gamma_n, \otimes_{n=1}^{\infty} \mathcal{A}_n, \prod_{n=1}^{\infty} \mathbb{P}_n\Big)$$

let us define the domains

$$\begin{aligned}\mathbb{D}_{(N)} = \Big\{ f \in L^2(\mathbb{P}) \colon \forall N,\ \text{for} \prod_{k=N+1}^{\infty} \mathbb{P}_k\text{-a.e. } \big(w_{N+1}, w_{N+2}, \dots\big) \\ \text{the function } f\big(\cdot, \dots, \cdot, w_{N+1}, w_{N+2}, \dots\big) \in \mathbb{D}^{(N)} \\ \text{and} \int \big(\Gamma_1[f] + \cdots + \Gamma_N[f]\big)\, d\mathbb{P} < +\infty \Big\}\end{aligned}$$

and for $f \in \mathbb{D}_{(N)}$ let us set

$$\Gamma_{(N)}[f] = \Gamma^{(N)}[f] = \Gamma_1[f] + \cdots + \Gamma_N[f].$$

It is easily seen that the terms $\big(\Omega, \mathcal{A}, \mathbb{P}, \mathbb{D}_{(N)}, \Gamma_{(N)}\big)$ are error structures. We remark that $\mathbb{D}_{(N)} \supset \mathbb{D}_{(N+1)}$ and if $f \in \mathbb{D}_{(N+1)}$ $\Gamma_{(N)}[f] \le \Gamma_{(N+1)}[f]$.

Let be

$$\mathbb{D} = \Big\{ f \in L^2(\mathbb{P}) \colon f \in \bigcap_N \mathbb{D}_{(N)},\ \lim_N \uparrow \Gamma_{(N)}[f] \in L^1(\mathbb{R}) \Big\}$$

and for $f \in \mathbb{D}$ let us put $\Gamma[f] = \lim_N \uparrow \Gamma_{(N)}[f]$.

In order to apply Lemma IV.7, it remains to be proved that $\mathbb{D}$ is dense in $L^2(\mathbb{P})$. This comes from the fact that $\mathbb{D}$ contains the cylindrical function f, such that f belongs to some $\mathbb{D}^{(N)}$.

The lemma provides the following theorem. The explicit definition of the domain $\mathbb{D}$ it gives for the product structure is particularly useful:

Theorem IV.8. *Let* $S_n = \big(\Omega_n, \mathcal{A}_n, \mathbb{P}_n, \mathbb{D}_n \Gamma_n\big)$, $n \geq 1$, *be error structures. The product structure*

$$S = (\Omega, \mathcal{A}, \mathbb{P}, \mathbb{D}, \Gamma) = \prod_{n=1}^{\infty} S_n$$

is defined by

$$(\Omega, \mathcal{A}, \mathbb{P}) = \Big(\prod_{n=1}^{\infty} \Omega_n, \otimes_{n=1}^{\infty} \mathcal{A}_n, \prod_{n=1}^{\infty} \mathbb{P}_n\Big)$$

$$\mathbb{D} = \Big\{ f \in L^2(\mathbb{P}) \colon \forall n, \textit{ for almost every } w_1, w_2, \dots, w_{n-1}, w_{n+1}, \dots \textit{ for the product measure } x \to f\big(w_1, \dots, w_{n-1}, x, w_{n+1}, \dots\big) \in \mathbb{D}_n \textit{ and } \int \sum_n \Gamma_n[f]\, d\mathbb{P} < +\infty \Big\}$$

and for $f \in \mathbb{D}$

$$\Gamma[f] = \sum_{n=1}^{\infty} \Gamma_n[f].$$

S is an error structure, Markovian if each S_n is Markovian.

As before, when we write $\Gamma_n[f]$, Γ_n acts on the n-th argument of f uniquely.

Let us add a comment about *projective systems*. The notion of projective systems of error structures can be defined similarly as in probability theory. Besides the topological assumptions (existence of a compact class, see Neveu [1964]) used in probability theory to ensure the existence of a limit, a new phenomenon appears whereby projective systems of error structures may have no (closed) limit. (See Bouleau–Hirsch, Example 2.3.4, Chapter V, p. 207 and Bouleau [2001].)

Nevertheless, when a projective system of error structures consists of images of a single error structure the projective limit does exist. Let us, for example, return to the case of the real-valued Markov chain recalled at the beginning of this chapter

$$X_{n+1} = \Psi\big(X_n, U_{n+1}\big) \quad X_0 = x.$$

If we suppose the U_n's to be independent with independent errors and considered as the coordinate mappings of the product error structure

$$S = \left([0,1], \mathcal{B}[0,1], dx, H^1([0,1]), \left(\frac{d}{dx}\right)^2\right)^{\mathbb{N}^*}$$

then the k-uples $\big(X_{n_1}, \dots, X_{n_k}\big)$ define a projective system that possesses a limit. It is an error structure on $\mathbb{R}^{\mathbb{N}^*}$ which is the image of S by the X_n's, in the sense of Proposition IV.3.

Exercise IV.9. Suppose $\Psi \in \mathcal{C}^1 \cap \text{Lip}$ and $\psi_2'(x,y)$ do not vanish. Show using functional calculus and Theorem III.12 that the pair $\big(X_n, \Gamma[X_n]\big)$ is a homogeneous Markov chain and that X_n has a density.

When we are interested in the existence of densities, as in the preceding exercise, the following proposition is useful for proving that Proposition III.16 is valid for an infinite product structure.

Proposition IV.10. *Consider a product error structure*

$$S = (\Omega, \mathcal{A}, \mathbb{P}, \mathbb{D}, \Gamma) = \prod_{n=1}^{\infty} S_n = \prod_{n=1}^{\infty} \big(\Omega_n, \mathcal{A}_n, \mathbb{P}_n, \mathbb{D}_n, \Gamma_n\big).$$

If every finite product of the S_n's satisfies Proposition III.16, *then S also satisfies this proposition: i.e.* $\forall u \in \mathbb{D}^k$

$$u_* \big(\det\big(\Gamma[u_i, u_j]\big) \cdot \mathbb{P}\big) \ll \lambda_k$$

where λ_k is the Lebesgue measure on $\mathbb{R}^k$.

Proof. The matrix

$$\underline{\underline{\Gamma}}\big[u, u^t\big] = \big(\Gamma[u_i, u_j]\big)_{ij}$$

is the increasing limit (in the sense of the order of positive symmetric matrices) of the matrices

$$\underline{\underline{\Gamma}}_N\big[u, u^t\big] = \big(\Gamma_n[u_i, u_j]\big)_{ij}$$

where Γ_N is defined as in the preparation of the theorem of products. Thus, if B is a Lebesgue-negligible set in $\mathbb{R}^k$

$$\int 1_B(u) \det \underline{\underline{\Gamma}}\big[u, u^t\big]\, d\mathbb{P} = \lim_N \int 1_B(u) \det \underline{\underline{\Gamma}}_N\big[u, u^t\big]\, d\mathbb{P} = 0. \qquad \diamond$$

Exercise IV.11.

1. Let D be a bounded connected open set in $\mathbb{R}^d$ with volume V. Show that the error structure

$$\left(D, \mathcal{B}(D), \frac{\lambda_d}{V}, H_0^1(D), \sum_{i=1}^{d} \left(\frac{\partial}{\partial x_i}\right)^2\right)$$

where λ_d is the Lebesgue measure satisfies the following inequality (Poincaré inequality)

$$\forall u \in H_0^1(D) \quad \|u\|_{L^2}^2 \leq k\mathcal{E}[u]$$

where the constant k can be taken to be $k = \frac{\ell^2}{d}$, with ℓ being the diameter of D. [*Hint*: for $u \in \mathcal{C}_K^\infty(D)$ integration along a parallel to the x_i-axis yields

$$u(x)^2 \le (x_i - x_i^*) \int \left(\frac{\partial u}{\partial x_i}\right)^2 dx_i$$

$$u(x)^2 \le (x_i^{**} - x_i) \int \left(\frac{\partial u}{\partial x_i}\right)^2 dx_i$$

where x_i^* is the infimum and x_i^{**} the supremum of D on this line.]

2. a) Let $f \in H^1([0,1])$, such that $f(0) = 0$. Using $f(x) = \int_0^x f'(t)\,dt$ and $2f'(s)f'(t) \le f'^2(s) + f'^2(t)$, show the inequality

$$\int_0^1 f^2(x)\,dx \le \int_0^1 f'^2(t)\frac{1-t^2}{2}\,dt.$$

   b) Deduce that the error structure

$$\left([0,1], \mathcal{B}([0,1]), dx, H^1([0,1]), \left(\frac{d}{dx}\right)^2\right)$$

   satisfies $\forall u \in H^1([0,1])$

$$\operatorname{var}[u] \le \frac{1}{2}\int_0^1 u'^2(x)\,dx = \mathcal{E}[u].$$

3. Let S be a product error structure

$$S = (\Omega, \mathcal{A}, \mathbb{P}, \mathbb{D}, \Gamma) = \prod_{n=1}^\infty (\Omega_n, \mathcal{A}_n, \mathbb{P}_n, \mathbb{D}_n, \Gamma_n) = \prod_{n=1}^\infty S_n.$$

Suppose that on each factor S_n holds an inequality

$$\forall u \in \mathbb{D}_n \quad \operatorname{var}[u] \le k\mathcal{E}_n[u].$$

   a) Show that for any $f \in L^2(\mathbb{P})$

$$\operatorname{var}_{\mathbb{P}}[f] \le \sum_{n=1}^\infty \mathbb{E}\operatorname{var}_{\mathbb{P}_n}[f].$$

   b) Deduce that $\forall f \in \mathbb{D}$

$$\operatorname{var}_{\mathbb{P}}[f] \le k\mathcal{E}[f].$$

4. Let $S = (\Omega, \mathcal{A}, \mathbb{P}, \mathbb{D}, \Gamma)$ be an error structure satisfying

$$\forall f \in \mathbb{D} \quad \operatorname{var}[f] \le k\mathcal{E}[f].$$

a) Show that any image of S satisfies the same inequality with the same k.

b) Admitting that the Ornstein–Uhlenbeck structure

$$S_{ou} = \left(\mathbb{R}, \mathcal{B}(\mathbb{R}), \mathcal{N}(0,1), H^1(\mathcal{N}(0,1)), \left(\frac{d}{dx}\right)^2\right)$$

satisfies the inequality with the best constant $k = 2$ (see Chafaï [2000]) show that S_{ou} is not an image of the Monte Carlo standard error structure

$$\left([0,1], \mathcal{B}[0,1], dx, H^1([0,1]), \left(\frac{d}{dx}\right)^2\right)^{\mathbb{N}}.$$

Appendix. Comments on projective limits

A projective system of error structures is a projective system of probability spaces on which quadratic error operators are defined in a compatible way.

Even when the probabilistic structures do possess a projective limit (which involves topological properties expressing roughly that the probability measures are Radon), a projective system of error structures may have no limit. It defines always a pre-structure, however this pre-structure may be non-closable.

When the probabilistic system is a product of the form

$$(\Omega, \mathcal{A}, \mathbb{P})^{\mathbb{N}}$$

the coordinate maps X_n represent repeated experiments of the random variable X_0. In such a case, if the quadratic error operator of the projective system defines correlated errors on the finite products, it may happen that the limit pre-structure be closable or not, depending on the special analytic form of the quadratic error operator, cf. Bouleau [2001] for examples.

In the non-closable case we may have

$$\lim_{M,N\uparrow\infty} \Gamma\Big[\frac{1}{M}\sum_{m=1}^{M} h(X_m) - \frac{1}{N}\sum_{n=1}^{N} h(X_n)\Big] = 0, \quad \text{in } L^1,$$

although $\Gamma[\frac{1}{N}\sum_{n=1}^{N} h(X_n)]$ does not converge to 0 when $N \uparrow \infty$.

This mathematical situation is related to a concrete phenomenon often encountered when doing measurements: *the error permanency under averaging*. Poincaré, in his discussion of the ideas of Gauss (*Calcul des Probabilités*) at the end of the nineteenth century, emphasizes this apparent difficulty and propose an explanation.

We shall return to this question in Chapter VIII, Section 2.

Bibliography for Chapter IV

N. Bouleau, Calcul d'erreur complet Lipschitzien et formes de Dirichlet, *J. Math. Pures Appl.* **80** (9) (2001), 961–976.

N. Bouleau and F. Hirsch, *Dirichlet Forms and Analysis on Wiener Space*, Walter de Gruyter, 1991.

D. Chafaï, L'exemple des lois de Bernoulli et de Gauss, in: *Sur les inégalités de Sobolev Logarithmiques*, Panor. Synthèses n°10, Soc. Math. France, 2000.

M. Fukushima, Y. Oshima and M. Takeda, *Dirichlet forms and Markov processes*, Walter de Gruyter, 1994.

J. Neveu, *Bases Mathématiques du Calcul des Probabilités*, Masson, 1964.

L. Schwartz, *Théorie des distributions*, Hermann, 1966.

Chapter V

Sensitivity analysis and error calculus

Thanks to the tools developed in the preceding chapter, we will explore some case studies for explaining the error calculus method.

We will start by taking simple examples from physics and finance.

Next, we will define a mathematical linearization of the quadratic operator Γ: the gradient. Several gradients are available which are isomorphic. The sharp (#) is a special choice of gradient, especially useful in stochastic calculus.

We will then present several integration by parts formulae, which are valid in any error structure. In the case of the Wiener space equipped with the Ornstein–Uhlenbeck structure, these formulae were the illuminating idea of Paul Malliavin in his works on improving classical results for stochastic differential equations (see Malliavin [1997]).

Afterwards, we will consider the case of an ordinary differential equation $y' = f(x, y)$ and determine the sensitivity of the solution to the infinite dimensional data f. Several approaches will be proposed.

We will conclude the chapter by examining the notion of error substructure which is the analog of a sub-σ-field for a probability space and the question of the projection on an error sub-structure which extends the notion of the conditional expectation.

1 Simple examples and comments

1.1 Cathodic tube. An oscillograph is modeled in the following way. After acceleration by an electric field, electrons arrive at point O_1 at a speed $v_0 > 0$ orthogonal to plane P_1. Between parallel planes P_1 and P_2, a magnetic field $\vec{B}$ orthogonal to O_1O_2 is acting; its components on O_2x and O_2y are (B_1, B_2).

a) Equation of the model. The physics of the problem is classical. The gravity force is negligible, the Lorenz force $q\vec{v} \wedge \vec{B}$ is orthogonal to $\vec{v}$ such that the modulus $|\vec{v}|$ remains constant and equal to v_0, and the electrons describe a circle of radius $R = \frac{mv_0}{e|\vec{B}|}$.

If θ is the angle of the trajectory with O_1O_2 as it passes through P_2, we then have

$$\theta = \arcsin \frac{a}{R}$$
$$|O_2A| = R(1 - \cos\theta)$$

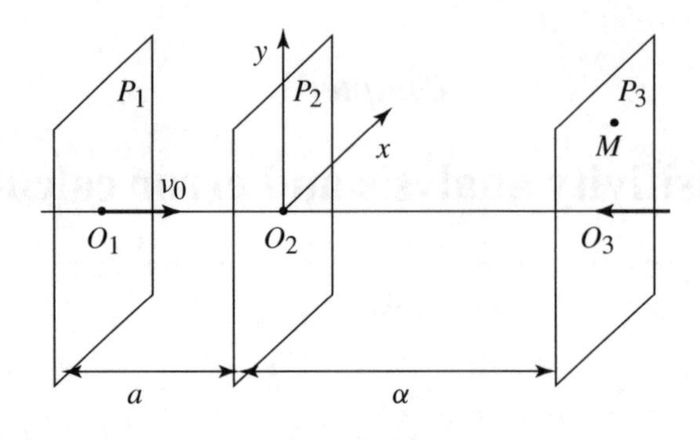

The trajectory of the electrons is incurved by $\vec{B}$ and the electrons finally reach the screen (plane P_3) where they illuminate point M.

and

$$A = \left(|O_2A| \frac{B_2}{|\vec{B}|}, -|O_2A| \frac{B_1}{|\vec{B}|} \right).$$

The position of M is thus given by

$$(1) \quad \begin{cases} M = (X, Y) \\ X = \left(\dfrac{mv_0}{e|\vec{B}|}(1 - \cos\theta) + d\tan\theta \right) \dfrac{B_2}{|\vec{B}|} \\ Y = -\left(\dfrac{mv_0}{e|\vec{B}|}(1 - \cos\theta) + d\tan\theta \right) \dfrac{B_1}{|\vec{B}|} \\ \theta = \arcsin \dfrac{ae}{mv_0} |\vec{B}| \end{cases}$$

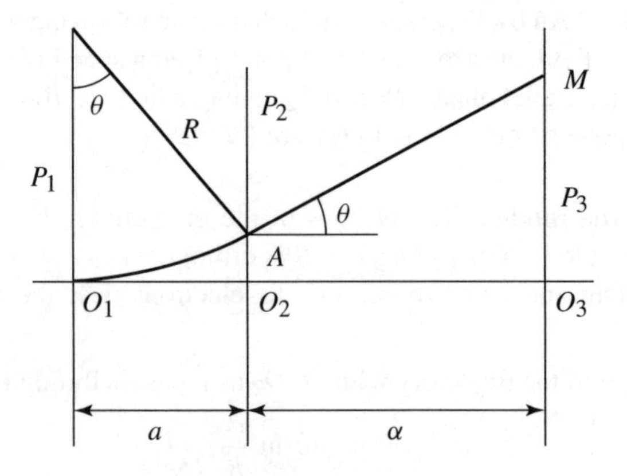

Figure in the plane of the trajectory

Numerical data:

$$\begin{aligned} m &= 0.1\ 10^{-31}\ \text{kg} \\ v_0 &= 2.3\ 10^{7}\ \text{ms}^{-1} \\ a &= 0.02\ \text{m} \\ d &= 0.27\ \text{m} \\ e &= 1.6\ 10^{-19}\ \text{C}. \end{aligned}$$

b) Sensitivity to $\vec{B}$. To study the sensitivity of point M to the magnetic field $\vec{B}$, we assume that $\vec{B}$ varies in $[-\lambda, \lambda]^2$ (with $\lambda = 4\ 10^{-3}$ Tesla), equipped with the error structure

$$\Big([-\lambda, \lambda]^2, \mathcal{B}\big([-\lambda, \lambda]^2\big), \mathbb{P}, \mathbb{D}, u \to {u'_1}^2 + {u'_2}^2\Big).$$

We can now compute the quadratic error on M, i.e. the matrix

$$\underline{\underline{\Gamma}}[M, M^t] = \begin{pmatrix} \Gamma[X] & \Gamma[X, Y] \\ \Gamma[X, Y] & \Gamma[Y] \end{pmatrix} = \begin{pmatrix} (\frac{\partial X}{\partial B_1})^2 + (\frac{\partial X}{\partial B_2})^2 & \frac{\partial X}{\partial B_1}\frac{\partial Y}{\partial B_1} + \frac{\partial X}{\partial B_2}\frac{\partial Y}{\partial B_2} \\ \frac{\partial X}{\partial B_1}\frac{\partial Y}{\partial B_1} + \frac{\partial X}{\partial B_2}\frac{\partial Y}{\partial B_2} & (\frac{\partial Y}{\partial B_1})^2 + (\frac{\partial Y}{\partial B_2})^2 \end{pmatrix}.$$

This computation is made possible using a symbolic calculus program. The result will not be presented herein due to space constraints.

However, if we were to simplify formulae (1) to the second order for small $|\vec{B}|$, we would obtain

$$\theta = \frac{ae}{mv_0}|\vec{B}| + \frac{1}{6}\left(\frac{ae}{mv_0}\right)^2 |\vec{B}|^3$$

and

$$X = \left[\left(\frac{a}{2} + d\right)\frac{ae}{mv_0} + \left(\frac{a}{12} + \frac{d}{2}\right)\left(\frac{ae}{mv_0}\right)^3 |\vec{B}|^2\right] B_2$$

$$Y = -\left[\left(\frac{a}{2} + d\right)\frac{ae}{mv_0} + \left(\frac{a}{12} + \frac{d}{2}\right)\left(\frac{ae}{mv_0}\right)^3 |\vec{B}|^2\right] B_1.$$

Thus

$$(2) \qquad \begin{cases} X = \big(\alpha + \beta\big(B_1^2 + B_2^2\big)\big)B_2 \\ Y = -\big(\alpha + \beta\big(B_1^2 + B_2^2\big)\big)B_1 \end{cases}$$

and we have

$$\underline{\underline{\Gamma}}[M, M^t]$$

$$(3) \quad = \begin{pmatrix} 4\beta^2 B_1^2 B_2^2 + \big(\alpha + \beta B_1^2 + 3\beta B_2^2\big)^2 & -2\beta B_1 B_2\big(\alpha + 3\beta\big(B_1^2 + B_2^2\big)\big) \\ -2\beta B_1 B_2\big(\alpha + 3\beta\big(B_1^2 + B_2^2\big)\big) & 4\beta^2 B_1^2 B_2^2 + \big(\alpha + \beta B_2^2 + 3\beta B_1^2\big)^2 \end{pmatrix}.$$

It follows in particular, by computing the determinant, that the law of M is absolutely continuous (see Proposition III.16 of Chapter III).

If we now suppose that the inaccuracy on the magnetic field stems from a noise in the electric circuit responsible for generating $\vec{B}$ and that this noise is *centered*: $A\big[B_1\big] = A\big[B_2\big] = 0$, we can compute the *bias* of the errors on $M = (X, Y)$:

$$A[X] = \frac{1}{2}\frac{\partial^2 X}{\partial B_1^2}\Gamma\big[B_1\big] + \frac{1}{2}\frac{\partial^2 X}{\partial B_2^2}\Gamma\big[B_2\big]$$
$$A[Y] = \frac{1}{2}\frac{\partial^2 Y}{\partial B_1^2}\Gamma\big[B_1\big] + \frac{1}{2}\frac{\partial^2 Y}{\partial B_2^2}\Gamma\big[B_2\big]$$

which yields

$$A[X] = 4\beta B_2$$
$$A[Y] = -4\beta B_1.$$

By comparison with (2), we can observe that

$$A\big[\overrightarrow{O_3M}\big] = \frac{4\beta}{\alpha + \beta|\vec{B}|^2}\,\overrightarrow{O_3M}.$$

In other words, the centered fluctuations of size ε^2 of the magnetic field induce not only an error on the spot, which is a small elliptical blotch described by matrix (3), but also a bias in the direction $\overrightarrow{O_3M}$ due to nonlinearities and (in the second-order approximation of the equation) equal to $\frac{4\beta}{\alpha+\beta|\vec{B}|^2}\,\overrightarrow{O_3M}\,\varepsilon^2$.

With the above numerical data at the extreme border of the screen where the error is largest, we obtain for the spot:

- a standard deviation in the direction $\overrightarrow{O_3M}$

$$\sigma_1 = 51\ \varepsilon,$$

- a standard deviation in the orthogonal direction

$$\sigma_2 = 45\ \varepsilon,$$

- a bias in the direction $\overrightarrow{O_3M}$

$$b = 16\ 10^3\ \varepsilon^2.$$

Taking $\varepsilon = 10^{-4}$ Tesla, which is a very large error, the standard deviation (about half a centimeter at the border of the screen) is 30 times greater than the bias. With a smaller error, the bias becomes negligible with respect to standard deviation.

In the following figure, we have changed α and β in order for the standard deviation and bias to both appear (the ellipse has the same meaning as in Chapter I, Section 2.2; it is a level curve of a very concentrated Gaussian density with $\Gamma[M, M^t]\varepsilon^2$ as covariance matrix and $A[M]\varepsilon^2$ as bias).

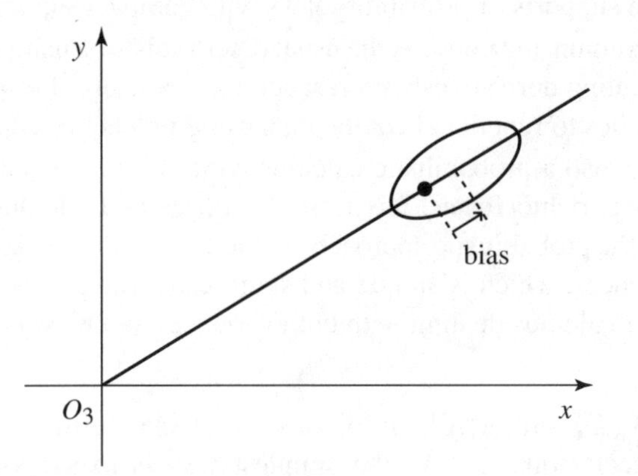

The appearance of biases in the absence of bias on the hypotheses is specifically due to the fact that the method considers the errors, although infinitesimal, to be *random* quantities. This feature will be highlighted in the following table.

1.2 Comparison of approaches. Let us provide an overview of the various approaches to error calculus.

Table. Main classes of error calculi

<table>
<tr><td rowspan="1">Deterministic approaches</td><td colspan="2">Deterministic sensitivity analysis:
derivation with respect to the parameters of the model</td><td>Interval calculus</td></tr>
<tr><td rowspan="2">Probabilistic approaches</td><td colspan="2">Error calculus using Dirichlet forms</td><td rowspan="2">Probability theory</td></tr>
<tr><td>first order calculus only dealing with variances</td><td>second order calculus with variances and biases</td></tr>
<tr><td></td><td colspan="2">Infinitesimal errors</td><td>Finite errors</td></tr>
</table>

At the right-hand side of the table and below, the usual probability calculus is displayed in which the errors are random variables. Knowledge of the joint laws of quantities and their errors is assumed to be yielded by statistical methods. The errors are finite and the propagation of the errors requires computation of image probability laws.

At the right-hand side and above is the interval calculus which, in some sense, is a calculus for the supports of probability laws with compact supports.

In the other column and above is the usual deterministic sensitivity calculus which consists of computing derivatives with respect to parameters. Let us remark that this calculus also applies to functional coefficients using Fréchet or Gâteaux derivatives.

Now, there is also a probabilistic calculus with infinitely small errors: the error calculus based on Dirichlet forms. It is a specific differential calculus taking in account some features of the probabilistic approach. In the same framework, either a first-order calculus on variances, which is simple and significant enough for most applications, or a second-order calculus dealing with both variances and biases can be performed.

1.3 Choice of $\Gamma[X]$ and $A[X]$. In order to implement error calculus on a model containing the scalar quantity X, the simplest way is to suppose that X and its error are independent of the remainder of the model, i.e. they take place by a factor $\big(\mathbb{R}, \mathcal{B}(\mathbb{R}), m, \boldsymbol{d}, \gamma\big)$, such that the error structure will have the form:

$$(\Omega, \mathcal{A}, \mathbb{P}, \mathbb{D}, \Gamma) = \big(\tilde{\Omega}, \tilde{\mathcal{A}}, \tilde{\mathbb{P}}, \tilde{\mathbb{D}}, \tilde{\Gamma}\big) \times \big(\mathbb{R}, \mathcal{B}(\mathbb{R}), m, \boldsymbol{d}, \gamma\big).$$

In such a situation, under typical regularity assumptions, we can choose $\Gamma[X]$ and $A[X]$ as we want (or at least the choice may be made with great freedom) and then study the propagation of both the variance and bias of error through the computation.

The variances of errors and the biases on regular functions will follow by means of

$$\begin{aligned} \Gamma\big[F(X)\big] &= F'^2(X)\Gamma[X] \\ A\big[F(X)\big] &= F'(X)A[X] + \frac{1}{2}F''(X)\Gamma[X], \end{aligned}$$

and if Y is defined on $\big(\tilde{\Omega}, \tilde{\mathcal{A}}, \tilde{\mathbb{P}}, \tilde{\mathbb{D}}, \tilde{\Gamma}\big)$ for regular G we will obtain

$$\begin{aligned} \Gamma\big[G(X,Y)\big] &= G_1'^2(X,Y)\Gamma[X] + G_2'^2(X,Y)\Gamma[Y] \\ A\big[G(X,Y)\big] &= G_1'(X,Y)A[X] + G_2'(X,Y)A[Y] + \frac{1}{2}G_{11}''(X,Y)\Gamma[X] \\ &\quad + \frac{1}{2}G_{22}''(X,Y)\Gamma[Y]. \end{aligned}$$

The measure m follows from the choice of $\Gamma[X]$ and $A[X]$, as shown by the following lemma, whose hypotheses could be weakened if we were to work in $\mathbb{D}_{\mathrm{loc}}$ (and in $\mathcal{D}A_{\mathrm{loc}}$ in a sense to be specified).

Lemma. *Let I be the identity map from $\mathbb{R} \to \mathbb{R}$. Let us impose $\Gamma[I]$ and $A[I]$. More formally, let g and a be two functions from $\mathbb{R}$ to $\mathbb{R}$ s.t. $g > 0$, with a and g being continuous. Let*

$$f(x) = \exp\left(\int_0^x \frac{2a(t)}{g(t)}\,dt\right).$$

We assume

$$(1+I^2)\frac{f}{g} \in L^1(\mathbb{R}, dx) \quad \text{and} \quad (1+I^2)f \in L^1(\mathbb{R}, dx).$$

Let k be the constant $\int_{\mathbb{R}} \frac{f(x)}{g(x)}\,dx$. Then the error pre-structure

$$\left(\mathbb{R}, \mathcal{B}(\mathbb{R}), \frac{f(x)}{kg(x)}\,dx, \mathcal{C}_K^1, u \to u'^2 g\right)$$

is closable such that $I \in \mathbb{D}$, $\Gamma[I] = g$ and $I \in \mathcal{D}A$, $A[I] = a$.

With $g = 1$ and $a(x) = -\frac{x}{a}$, we obtain the Ornstein–Uhlenbeck structure.

Proof. a) The function $\frac{f}{kg}$ and g are continuous hence, by means of the Hamza condition, the pre-structure is closable.

b) Let us show that $\mathcal{C}_K^1$ is dense in $\mathcal{C}^1 \cap \mathrm{Lip}$ for the $\mathbb{D}$-norm.

Let $u \in \mathcal{C}^1 \cap \mathrm{Lip}$ with Lipschitz constant C. Let $\varphi_n \in \mathcal{C}_K^1$ s.t. $\varphi_n(x) = 1$ on $[-n, n]$ and $\varphi_n(x) = 0$ outside $]-n-2, n+2[$, $\|\varphi_n'\|_\infty \le 1$, $|\varphi_n| \le 1$. The $\mathcal{C}_K^1$-functions $\varphi_n \cdot u$ then verify

$$\begin{aligned}
\varphi_n \cdot u &\to u \quad \text{a.e.}\\
\varphi_n' u + \varphi_n u' &\to u' \quad \text{a.e.}\\
|\varphi_n u| &\le C|I| + C_1 \in L^2\left(\frac{f}{g}\,dx\right)\\
|\varphi_n' u + \varphi_n u'| &\le C|I| + C_1 + C \in L^2(f\,dx)
\end{aligned}$$

which yields the result by dominated convergence.

c) Hence $I \in \mathbb{D}$ and by the above approximation

$$\Gamma[I] = g.$$

d) Let us use the lemma of Remark III.5 to show $I \in \mathcal{D}A$:

$$\begin{aligned}
\mathcal{E}[I, u] &= \frac{1}{2}\int u'(x)g(x)\frac{f(x)}{kg(x)}\,dx\\
&= -\langle u, a\rangle_{L^2\left(\frac{f}{kg}\,dx\right)} \quad \forall u \in \mathcal{C}_K^1.
\end{aligned}$$

Thus, $I \in \mathcal{D}A$ and

$$A[I] = a. \qquad \diamond$$

Remark V.1 (No error on the constants!). In an error structure the measure $\mathbb{P}$ is compulsory, although in some almost-sure calculations with Γ and A, it may be unspecified as explained above.

It is not possible, within this theory, to consider a factor $\big(\mathbb{R}, \mathcal{B}(\mathbb{R}), m, \boldsymbol{d}, \gamma\big)$ in which m would be a Dirac mass, unless γ were zero. In other words, *erroneous quantities are necessarily random.* The theory does not address the question of either the error on π or the error on the speed of light. For this kind of situation, when investigating the propagation of inaccuracy, we must rely on ordinary differential calculus or, if focusing on covariances and biases, content ourselves with error pre-structures.

The present theory focuses more on the measurement device than on the quantity itself. As Gauss tackled the problem, the quantity may vary inside the scope of the instrument (here is the probability law) and the measurement yields a result and an error both depending on the physical quantity. Thus, a Dirac mass would correspond to a measurement device only able to measure one single value!

Example V.2 (Interest rate models). Consider a financial model of interest rate. The short-term rate $r(t)$ is a stochastic process and the price at time t of a discount bond with principal 1 Euro, (the so-called "zero-coupon" bond), can be computed, with some additional hypotheses, thanks to the formula

$$P(t,T) = \mathbb{E}\left[e^{-\int_t^T r(s)\,ds} \mid \mathcal{F}_t\right].$$

Once the model parameters have been calibrated by use of market data, the error calculus can assess the sensitivity of $r(0)$ and $P(0,T)$ to the hypotheses. It is also natural to be interested in quantities $r(h)$ and $P(h,T)$ for $h > 0$ as well as in their best estimates provided by the model at time 0.

a) Elementary model. Suppose that the short-term rate can be written as follows:

$$r(t) = b + a_1 e^{-t} + a_2 t e^{-t} + a_3 \frac{t^2}{2} e^{-t}$$

with a_1, a_2 and a_3 being random, independent and uniformly distributed on $[-b, b]$. Necessarily

$$\begin{aligned} a_1 &= r(0) - b \\ a_2 &= r'(0) + r(0) - b \\ a_3 &= r''(0) + 2r'(0) + r(0) - b \end{aligned}$$

and the behavior of $r(t)$ is deterministic after $t = 0$.

$$\begin{aligned} P(0,T) = \exp\bigg\{ -\Big(& bT + a_1(1 - e^{-T}) + a_2(1 - (1+T)e^{-T}) \\ & + a_3\Big(1 - \Big(1 + T + \frac{T^2}{2}\Big)e^{-T}\Big)\Big)\bigg\}. \end{aligned}$$

Supposing a_1, a_2 and a_3 are erroneous and modeled on

$$\left([b,b], \mathcal{B}, \frac{dx}{2b}, H^1[[-b,b]), u \to u'^2\right)^3,$$

we obtain that the proportional quadratic error on the zero-coupon bond is deterministic:

$$\frac{\Gamma[P(0,T)]}{(P(0,T))^2} = \left(1-e^{-T}\right)^2 + \left(1-(1+T)e^{-T}\right)^2 + \left(1-\left(1+T+\frac{T^2}{2}\right)e^{-t}\right)^2.$$

This example displays clearly the mathematical difficulties encountered when speaking about biases: intuitively by the symmetry of the hypotheses $r(t)$ has no bias since $A[a_1] = A[a_2] = A[a_3] = 0$. Instead $P(0,T)$ is biased and its bias is always strictly positive:

$$\frac{A[P(0,T)]}{P(0,T)} = \frac{1}{2}\frac{\Gamma[P(0,T)]}{(P(0,T))^2}.$$

However, this is not completely correct by the fact that a_1 does not belong to the domain of the generator in L^2-sense. If the domain of the form is taken to be $H^1[-b,b]$, functions in $\mathcal{D}A$ must satisfy $f(b) = f(-b) = 0$ as shown, once again, by the lemma of Remark III.5.

We shall not investigate deeper this field of research in the present book. It would be necessary to introduce the notion of extended generator (cf. H. Kunita [1969], P.-A. Meyer [1976], G. Mokobodzki [1978], D. Feyel [1978], N. Bouleau [1981]).

b) Vasicek model. It is supposed that

$$r(t) - b = z(t)$$

is an Ornstein–Uhlenbeck process, i.e. a centered Gaussian process that is Markovian with semigroup (see Chapter II):

$$P_t F(x) = \int F\left(\sqrt{e^{-\alpha t}}\, x + \sqrt{1-e^{-\alpha t}}\, \beta y\right) dm(y)$$

where $m = \mathcal{N}(0,1)$ is the reduced normal law and α, β are two strictly positive parameters.

The best estimate of $r(h)$ given $\mathcal{F}_0$ is usually obtained using the Wiener filtering procedure, yet is highly simplified herein by the Markovian property:

$$\mathbb{E}\left[r(h) \mid \mathcal{F}_0\right] = b + P_h[I](z_0) = b + \sqrt{e^{-\alpha h}}\,(r(0)-b) \tag{4}$$

where I denotes the identity map on $\mathbb{R}$.

The price of the zero-coupon bond is also computed thanks to the Markov property,

$$P(0,T) = \mathbb{E}\Big[e^{-\int_0^T r(s)\,ds} \mid \mathcal{F}_0\Big] = \mathbb{E}_{z_0}\Big[e^{-\int_0^T r(s)\,ds}\Big],$$

and using the fact that $\int_0^T r(s)\,ds$ is a Gaussian random variable with mean

$$\mathbb{E}_{z_0}\left[\int_0^T r(s)\,ds\right] = \frac{1-e^{-\frac{\alpha}{2}T}}{\frac{\alpha}{2}} z_0 + bT = M(\alpha,T)z_0 + bT$$

and with variance

$$\begin{aligned}\mathrm{Var}_{z_0}\left[\int_0^T r(s)\,ds\right] &= \frac{4\beta^2}{\alpha}\left[T - \frac{2}{\alpha}\left(1-e^{-\frac{\alpha}{2}T}\right) - \frac{1}{\alpha}\left(1-e^{-\frac{\alpha}{2}T}\right)^2\right]\\ &= V(\alpha,\beta,T)\end{aligned}$$

we then obtain

$$P(0,T) = e^{-M(\alpha,T)z_0 + \frac{1}{2}V(\alpha,\beta,T) - bT}. \tag{5}$$

It follows that

$$\begin{aligned}\mathbb{E}\big[P(h,T) \mid \mathcal{F}_0\big] &= \mathbb{E}\left[e^{-M(\alpha,T-h)z_h + \frac{1}{2}V(\alpha,\beta,T-h) - b(T-h)} \mid \mathcal{F}_0\right]\\ &= e^{\frac{1}{2}V(\alpha,\beta,T-h) - b(T-h)} \int e^{-M(\alpha,T-h)\left(\sqrt{e^{-\alpha h}}z_0 + \sqrt{1-e^{-\alpha h}}\beta y\right)}\,dm(y)\end{aligned}$$

and

$$\mathbb{E}\big[P(h,T) \mid \mathcal{F}_0\big] = \exp\left\{-Me^{-\frac{\alpha}{2}h}z_0 + \frac{1}{2}V - b(T-h) + \frac{1}{2}M^2\left(1-e^{-\alpha h}\right)\beta^2\right\}. \tag{6}$$

c) Alternative case. If we take

$$r(t) - b = y(t)$$

where $y(t)$ is the stationary Gaussian process solution to

$$y'(t) + y(t) = z(t)$$

where $z(t)$ is as stated above, $r(t)$ becomes a more regular process (with finite variation). The computation can be conducted similarly. Noting that

$$y(t) = e^{-t}y(0) + e^{-t}\int_0^t e^s z(s)\,ds$$

and using the Gaussian and Markov properties of $z(t)$, we obtain

$$\mathbb{E}\big[r(h) \mid \mathcal{F}_0\big] = b + e^{-h}y(0) + he^{-h}z_0 \quad \text{if } \alpha = 2 \tag{7}$$

$$= b + e^{-h}y(0) + \frac{e^{-\frac{\alpha}{2}h} - e^{-h}}{1-\frac{\alpha}{2}} z_0 \quad \text{if } \alpha \neq 2 \tag{8}$$

and

$$P(0,T) = e^{K+Ly(0)+Ny'(0)} = e^{K'+Lr(0)+Nr'(0)} \tag{9}$$

where the constants K, L, M, K' depend on b, α, β, T.

Formulae (4) to (9) allow studying the sensitivities and cross-sensitivities of quantities $r(0)$, $r(h)$, $P(0,T)$, $P(h,T)$, $\mathbb{E}[r(h) \mid \mathcal{F}_0]$ and $\mathbb{E}[P(h) \mid \mathcal{F}_0]$ with respect to parameters b, α, β, h and T, in the same spirit as that adopted in the case of the cathodic tube. These parameters must first be randomized and we obtain $\Gamma[P(0,T)]$ and $A[P(0,T)]$ for instance, as functions of the Γ and A of the parameters.

At this point, let us note that our model is different from a finite dimensional model by the presence of a *stochastic process*, here stationary, $z(t)$ and we may also consider this process as an erroneous hypothesis and identify the consequences of this error.

Suppose that the process $z(t)$ is defined on an error structure (see Chapter VI), such that $z(t) \in \mathbb{D}$ and even $z(t) \in \mathcal{D}A\ \forall t$, and furthermore

$$\Gamma[z(t+h), z(t)] = e^{-h}$$
$$A[z(t)] = -\frac{1}{2}z(t).$$

From the relation (5) we draw the following:

$$\begin{aligned}
\Gamma[P(0,T)] &= M^2(P(0,T))^2 \\
\Gamma[P(0,T), P(h,T)] &= M(\alpha,T)M(\alpha,T-h)P(0,T)P(h,T)e^{-h} \\
A[P(0,t)] &= \frac{1}{2}M(\alpha,T)P(0,T)z(0) + \frac{1}{2}\big(M(\alpha,T)\big)^2 P(0,T).
\end{aligned}$$

We now see that the proportional quadratic error

$$\frac{\Gamma[P(0,T)]}{\big(P(0,T)\big)^2} = \big(M(\alpha,T)\big)^2$$

is constant (does not depend on $r(0)$) and that the proportional bias

$$\frac{A[P(0,T)]}{P(0,T)} = \frac{1}{2}M(\alpha,T)z(0) + \frac{1}{2}\big(M(\alpha,T)\big)^2$$

is an affine function of $z(0) = r(0) - b$ with a positive value at the origin.

In a more complete study, it would be interesting to consider several error structures, since, as we will see in Chapter VI, they can yield appreciably different results for infinite dimensional models.

Remark V.3 (Best estimates and prediction). Regarding forecasting, i.e. finding the best estimate of $r(h)$ or $P(h,T)$, we can project $r(h)$ and $P(h,T)$ onto the space of what is known at time zero, either in the sense of L^2 or in the sense of the norm $\|\cdot\|_{\mathbb{D}}$, or even in the sense of another mixed norm

$$\big(\alpha\|\cdot\|_{L^2} + (1-\alpha)\mathcal{E}[\cdot]\big)^{1/2}.$$

Under hypotheses in which these projections exist, the result will in general be different. The L^2-projection does not take errors into account. The $\mathbb{D}$-norm or mixed norms provide more regular results (in $\mathbb{D}$).

We shall study this question in Section 4 of the present chapter.

2 The gradient and the sharp

One of the features of the operator Γ is to be quadratic or bilinear like the variance or covariance. That often makes computations awkward to perform. If we accept to consider random variables with values in Hilbert space, it is possible to overcome this problem by introducing a new operator, the gradient, which in some sense is a *linear version of the standard deviation of the error*.

2.1 The gradient. Let $S = (\Omega, \mathcal{A}, \mathbb{P}, \mathbb{D}, \Gamma)$ be an error structure. If $\mathcal{H}$ is a real Hilbert space, we denote either by $L^2\big((\Omega, \mathcal{A}, \mathbb{P}), \mathcal{H}\big)$ or $L^2(\mathbb{P}, \mathcal{H})$ the space of $\mathcal{H}$-valued random variables equipped with the scalar product

$$(U, V)_{L^2(\mathbb{P},\mathcal{H})} = \mathbb{E}\big[\langle U, V\rangle_{\mathcal{H}}\big].$$

This space can be identified with the Hilbertian (completed) tensor product $L^2(\Omega, \mathcal{A}, \mathbb{P}) \otimes \mathcal{H}$.

Definition V.4. *Let $\mathcal{H}$ be a Hilbert space. A linear operator D from $\mathbb{D}$ into $L^2(\mathbb{P}, \mathcal{H})$ is said to be a gradient (for S) if*

$$\forall u \in \mathbb{D} \quad \Gamma[u] = \langle Du, Du\rangle_{\mathcal{H}}.$$

In practice, a gradient always exists thanks to the following result that we admit herein (see Bouleau and Hirsch [1991] Exercise 5.9 p. 242).

Fact V.5 (G. Mokobodzki). Once the space $\mathbb{D}$ is separable (i.e. possesses a dense sequence), there exists a gradient for S.

Proposition V.6. *Let D be a gradient for S with values in $\mathcal{H}$. Then $\forall u \in \mathbb{D}^n$ $\forall F \in \mathcal{C}^1 \cap \mathrm{Lip}(\mathbb{R}^n)$,*

$$D[F \circ u] = \sum_{i=1}^{n} \frac{\partial F}{\partial x_i} \circ u \; D[u_i] \quad \text{a.e.}$$

Proof. Considering the difference of the two members, we have

$$\Big\| D[F \circ u] - \sum_i \frac{\partial F}{\partial x_i} \circ u \; D[u_i] \Big\|_{\mathcal{H}}^2 = \Gamma[F \circ u] - 2 \sum_i \frac{\partial F}{\partial x_i} \circ u \, \Gamma\big[F \circ u, u_i\big] + \sum_{ij} \frac{\partial F}{\partial x_i} \circ u \, \frac{\partial F}{\partial x_j} \circ u \, \Gamma\big[u_i, u_j\big].$$

This expression however vanishes by virtue of the functional calculus for Γ. $\diamond$

The same argument shows that Lipschitzian calculus on functions of one variable is also valid for D.

For the examples of finite dimensional structures given in Chapter III, a gradient is easily constructed by taking a finite dimensional Euclidean space for $\mathcal{H}$.

In the classical case of a domain of $\mathbb{R}^d$ with $\Gamma[u] = \sum_i \left(\frac{\partial u}{\partial x_i}\right)^2$ and the Lebesgue measure (Example III.6), the gradient coincides with the usual gradient on regular functions.

In the case of Example III.23 where $\Gamma[u] = \sum_{ij} a_{ij}(x) \frac{\partial u}{\partial x_i} \frac{\partial u}{\partial x_j}$, a gradient may be taken as $D[u] = B \cdot \nabla u$, where matrix B satisfies $B^t B = \big(a_{ij}\big)$.

Obtaining a gradient for a product is straightforward. To acknowledge this, let us introduce, for a sequence of Hilbert space $\mathcal{H}_n$, the direct sum

$$\mathcal{H} = \bigoplus_n \mathcal{H}_n$$

as a subvector space of $\prod_n \mathcal{H}_n$ equipped with the scalar product

$$\big\langle (u_1, \ldots, u_n, \ldots), (v_1, \ldots, v_n, \ldots) \big\rangle_{\mathcal{H}} = \sum_{n=1}^{\infty} \langle u_n, v_n \rangle_{\mathcal{H}_n}.$$

Proposition V.7. *Let $S_n = \big(\Omega_n, \mathcal{A}_n, \mathbb{P}_n, \mathbb{D}_n, \Gamma_n\big)$ be error structures equipped with gradients $\big(D_n, \mathcal{H}_n\big)$. Then the product*

$$S = (\Omega, \mathcal{A}, \mathbb{P}, \mathbb{D}, \Gamma) = \prod_n S_n$$

possesses a gradient with values in $\mathcal{H} = \bigoplus_n \mathcal{H}_n$, given for $u \in \mathbb{D}$ by

$$D[u] = \big(D_1[u], \ldots, D_n[u], \ldots\big)$$

where $D_n[u]$ means that the operator D_n acts on the n-th variable of u.

Proof. The demonstration follows straightforwardly from the theorem on products (Theorem IV.8) using $\Gamma[u] = \sum_n \Gamma_n[u]$. $\diamond$

The behavior of the gradient unfortunately is not very good by images. Taking the image of S by $X \in \mathbb{D}$, the natural candidate

$$D_X[u](x) = \mathbb{E}\big[D[u \circ X] \mid X = x\big]$$

is not a gradient for the image structure because

$$\mathbb{E}\big[\langle D[u\circ X], D[u\circ X]\rangle_{\mathcal{H}} \mid X = x\big] \\ \neq \big\langle \mathbb{E}\big[D[u\circ X] \mid X = x\big], \mathbb{E}\big[D[u\circ X] \mid X = x\big]\big\rangle_{\mathcal{H}}.$$

In practice, when considering image structures, we must reconstruct a new gradient on the new structure. In terms of errors, this remark means that the gradient is not intrinsic, but rather a derived concept dependent on the presentation we adopt.

2.2 The sharp. Let $(\Omega, \mathcal{A}, \mathbb{P}, \mathbb{D}, \Gamma)$ be an error structure and $(\widehat{\Omega}, \widehat{\mathcal{A}}, \widehat{\mathbb{P}})$ be a copy of $(\Omega, \mathcal{A}, \mathbb{P})$.

As soon as the operator Γ does not vanish identically, the space $L^2(\Omega, \mathcal{A}, \mathbb{P})$ is compulsory infinite dimensional. Indeed, let us suppose that the σ-field $\mathcal{A}$ is generated by a finite number of disjoint atoms A_i. Then any random variable is of the form $\sum a_i 1_{A_i}$ and the functional calculus implies $\Gamma[1_{A_i}] = 0$, hence $\Gamma = 0$.

Thus the copy $L^2(\widehat{\Omega}, \widehat{\mathcal{A}}, \widehat{\mathbb{P}})$ is infinite dimensional. Therefore, if $(D, \mathcal{H})$ is a gradient for S, there is an isometry J from $\mathcal{H}$ into $L^2(\widehat{\Omega}, \widehat{\mathcal{A}}, \widehat{\mathbb{P}})$ and setting $u^\# = J\big(D[u]\big)$ yields a gradient with values in $L^2(\widehat{\Omega}, \widehat{\mathcal{A}}, \widehat{\mathbb{P}})$ and $L^2\big(\mathbb{P}, L^2(\widehat{\Omega}, \widehat{\mathcal{A}}, \widehat{\mathbb{P}})\big)$ being identified with $L^2(\Omega \times \widehat{\Omega}, \mathcal{A} \times \widehat{\mathcal{A}}, \mathbb{P} \times \widehat{\mathbb{P}})$. We then obtain:

- $\forall u \in \mathbb{D} \quad u^\#(w, \widehat{w}) \in L^2(\mathbb{P} \times \widehat{\mathbb{P}})$
- $\forall u \in \mathbb{D} \quad \Gamma[u] = \widehat{\mathbb{E}}\big[(u^\#)^2\big]$
- $\forall u \in \mathbb{D}^n \quad F \in \mathcal{C}^1 \cap \mathrm{Lip}\big(\mathbb{R}^n\big) \ \big(F(u_1, \dots, u_n)\big)^\# = \sum_i \frac{\partial F}{\partial x_i} \circ u \cdot u_i^\#.$

This choice is particularly useful for stochastic calculus with error structures on the Wiener space (see Chapter VI) and also in ergodic theory (see Bouleau and Hirsch Chapter VII).

Remark. If in Proposition V.7 the gradient on each factor is in fact a sharp, then the gradient obtained on the product structure is also a sharp. This finding is due to the simple fact that the product of the copy spaces is a copy of the product.

2.3 The adjoint δ or "divergence". Since the domain $\mathbb{D}$ of D is dense in $L^2(\mathbb{P})$, the gradient D possesses an adjoint operator. This adjoint δ is defined on a domain included in $L^2(\mathbb{P}, \mathcal{H})$ with values in $L^2(\mathbb{P})$: in $L^2(\mathbb{P})$:

$$\operatorname{dom}\delta = \big\{U \in L^2(\mathbb{P}, \mathcal{H}) : \text{ there exists a constant } C \text{ such that} \\ |\mathbb{E}\langle D[u], U\rangle_{\mathcal{H}}| \le C\|u\|_{L^2(\mathbb{P})} \ \forall u \in \mathbb{D}\big\}$$

and if $U \in \operatorname{dom}\delta$

$$(10) \qquad \langle \delta U, u\rangle_{L^2(\mathbb{P})} = \langle U, D[u]\rangle_{L^2(\mathbb{P}, \mathcal{H})} \quad \forall u \in \mathbb{D}.$$

Adjoint operators are always closed.

3 Integration by parts formulae

Let $S = (\Omega, \mathcal{A}, \mathbb{P}, \mathbb{D}, \Gamma)$ be an error structure. If $v \in \mathcal{D}A$, for all $u \in \mathbb{D}$ we have

$$\frac{1}{2}\mathbb{E}\big[\Gamma[u,v]\big] = -\mathbb{E}\big[uA[v]\big]. \tag{11}$$

This relation is already an integration by parts formula since Γ follows first-order differential calculus, in particular if $F \in \mathrm{Lip}$ with Lebesgue derivative F':

$$\frac{1}{2}\mathbb{E}\big[F'(u)\Gamma[u,v]\big] = -\mathbb{E}\big[F(u)A[v]\big]. \tag{12}$$

We know that $\mathbb{D} \cap L^\infty$ is an algebra (Proposition III.10), hence if $u_1, u_2 \in \mathbb{D} \cap L^\infty$ we can apply (11) to u_1u_2 as follows:

$$\frac{1}{2}\mathbb{E}\big[u_2\Gamma[u_1,v]\big] = -\mathbb{E}\big[u_1u_2A[v]\big] - \frac{1}{2}\mathbb{E}\big[u_1\Gamma[u_2,v]\big] \tag{13}$$

which yields for φ Lipschitz

$$\frac{1}{2}\mathbb{E}\big[u_2\varphi'(u_1)\Gamma[u_1,v]\big] = -\mathbb{E}\big[\varphi(u_1)u_2A[v]\big] - \frac{1}{2}\mathbb{E}\big[\varphi(u_1)\Gamma[u_2,v]\big]. \tag{14}$$

Let us now introduce a gradient D with values in $\mathcal{H}$ along with its adjoint operator δ. The preceding formula (10) with $u \in \mathbb{D}$, $U \in \operatorname{dom}\delta$

$$\mathbb{E}[u\delta U] = \mathbb{E}\big[\langle D[u], U\rangle_{\mathcal{H}}\big] \tag{15}$$

provides, as above, for φ Lipschitz

$$\mathbb{E}\big[\varphi'(u)\langle D[u], U\rangle_{\mathcal{H}}\big] = \mathbb{E}[\varphi(u)\delta U]. \tag{16}$$

Moreover if $u_1, u_2 \in \mathbb{D} \cap L^\infty$ and $U \in \operatorname{dom}\delta$

$$\mathbb{E}\big[u_2\langle Du_1, U\rangle_{\mathcal{H}}\big] = \mathbb{E}\big[u_1u_2\delta U\big] - \mathbb{E}\big[u_1\langle Du_2, U\rangle_{\mathcal{H}}\big]. \tag{17}$$

Application: The internalization procedure (see Fournié et al. [1999]). Suppose the following quantity is to be computed

$$\frac{d}{dx}\mathbb{E}[F(x,w)].$$

If the space Ω can be factorized $\Omega = \mathbb{R} \times \tilde{\Omega}$ such that $w \in \Omega$ be represented $w = (w_1, w_2)$ with $w_1 \in \mathbb{R}$ and the random variable $F(x,w)$ be written

$$F(x,w) = \Phi\big(f(x,w_1), w_2\big)$$

we can consider w_1 to be erroneous and place an error structure on it, e.g. $\big(\mathbb{R}, \mathcal{B}(\mathbb{R}), \mathcal{N}(0,1), H^1(\mathcal{N}(0,1)), u \to u'^2\big)$ with gradient D (with $\mathcal{H} = \mathbb{R}$) and adjoint δ.

Supposing the necessary regularity assumptions to be fulfilled, we can now write

$$\frac{\partial F}{\partial x}(x, w) = \frac{\partial f}{\partial x}(x, w_1)\Phi_1'\big(f(x, w_1), w_2\big)$$

and

$$DF = \frac{\partial f}{\partial w_1}(x, w_1)\Phi_1'\big(f(x, w_1), w_2\big).$$

Thus

$$\frac{\partial F}{\partial x} = DF \cdot \frac{\partial f}{\partial x}\left(\frac{\partial f}{\partial w_1}\right)^{-1},$$

and by (17) with $U = 1$

$$\begin{aligned}
\mathbb{E}\frac{\partial F}{\partial x} &= \mathbb{E}\left[\frac{\partial f}{\partial x}\left(\frac{\partial f}{\partial w_1}\right)^{-1} DF\right] \\
&= \mathbb{E}\left[F\frac{\partial f}{\partial x}\left(\frac{\partial f}{\partial w_1}\right)^{-1} \delta[1]\right] - \mathbb{E}\left[FD\left[\frac{\partial f}{\partial x}\left(\frac{\partial f}{\partial w_1}\right)^{-1}\right]\right].
\end{aligned}$$

Finally

(18)
$$\frac{d}{dx}\mathbb{E}[F] = \mathbb{E}\left[\Phi\big(f(x, w_1), w_2\big)\left\{\frac{\partial f}{\partial x}\left(\frac{\partial f}{\partial w_1}\right)^{-1} \delta[1] - D\left[\frac{\partial f}{\partial x}\left(\frac{\partial f}{\partial w_1}\right)^{-1}\right]\right\}\right].$$

If the term $\{\cdot\}$ can be explicitely computed, the right-hand member is easier to obtain by Monte Carlo simulation than is the left-hand side a priori, especially when the function Φ often varies while f remains the same.

4 Sensitivity of the solution of an ODE to a functional coefficient

In this section, we are interested in the following topics.

(i) Error calculus on $f(X)$ when f and X are erroneous.

(ii) Error calculus for an ordinary differential equation $y' = f(x, y)$, $y(0) = Y_0$ when f, x and Y_0 are erroneous.

4.1 Error structures on functional spaces. Let Ω_1 be a space of functions (not necessarily a vector space), e.g. from $\mathbb{R}^d$ into $\mathbb{R}$, and let us consider an error structure $S_1 = \big(\Omega_1, \mathcal{A}_1, \mathbb{P}_1, \mathbb{D}_1, \Gamma_1\big)$ with the following properties.

a) The measure $\mathbb{P}_1$ is carried by the $\mathcal{C}^1 \cap \mathrm{Lip}$ function in Ω_1.

b) Let $S_2 = \big(\mathbb{R}^d, \mathcal{B}(\mathbb{R}^d), \mathbb{P}_2, \mathbb{D}_2, \Gamma_2\big)$ be an error structure on $\mathbb{R}^d$ such that $\mathcal{C}^1 \cap \mathrm{Lip} \subset \mathbb{D}_2$. Let us suppose the following: If we denote V_x the valuation at x, defined via
$$V_x(f) = f(x) \quad f \in \Omega_1,$$
V_x is a real functional on Ω_1 (a linear form if Ω_1 is a vector space). We now suppose that for $\mathbb{P}_2$-a.e. x we have $V_x \in \mathbb{D}_1$, and the random variable F defined on $\Omega_1 \times \mathbb{R}^d$ by
$$F(f, x) = V_x(f)$$
satisfies
$$\int \big(\Gamma_1[F] + \Gamma_2[F]\big)\, d\mathbb{P}_1\, d\mathbb{P}_2 < +\infty.$$

The theorem on products then applies and we can write

$$\Gamma[F] = \Gamma_1[F] + \Gamma_2[F]. \tag{19}$$

Let us consider that the two structures S_1 and S_2 have a sharp operator. This assumption gives rise to a sharp on the product structure (see the remark in Section 2.5).

Proposition V.8. *With the above hypotheses, yet with $d = 1$ for the sake of simplicity, let $X \in \mathbb{D}_2$, then $f(X) \in \mathbb{D}$ and*

$$\big(f(X)\big)^{\#} = f^{\#}(X) + f'(X)X^{\#}. \tag{20}$$

Proof. This system of notation has to be explained: On the left-hand side, $f(X)$ is a random variable on $\Omega_1 \times \mathbb{R}^d$, defined via
$$F(f, x) = V_x(f)$$
and
$$f(X) = F(f, X).$$
The first term on the right-hand side denotes the function of x
$$\big(V_x\big)^{\#} \quad \text{taken on } X.$$
The second term is clearly apparent, f' is the derivative of f (since f is $\mathcal{C}^1 \cap \mathrm{Lip}$ $\mathbb{P}_1$-a.s.). With this notation, the proposition is obvious from Proposition V.6. ◇

Examples V.9. We will construct examples of situations as described above when the space Ω_1 is

(j) either a space of analytic functions in the unit disk with real coefficients

$$f(z) = \sum_{n=0}^{\infty} a_n z^n$$

(jj) or a space of L^2 functions, e.g. in $L^2[0, 1]$

$$f(x) = \sum_{n\in\mathbb{Z}} a_n e^{2i\pi nx},$$

with, in both cases, setting the a_n to be random with a product structure on them.

4.2 Choice of the a priori probability measure. If we choose the a_n to be i.i.d., the measure $\mathbb{P}_1$ is carried by a very small set and the scaling $f \to \lambda f$ gives from $\mathbb{P}_1$ a singular measure.

In order for $\mathbb{P}_1$ to weigh on a cone or a vector space, we use the following result.

Property. *Let μ be a probability measure on $\mathbb{R}$ with a density ($\mu \ll dx$). We set*

$$\mu_n = \alpha_n \mu + \big(1 - \alpha_n\big)\delta_0$$

with $\alpha_n \in]0, 1[$, $\sum_n \alpha_n < +\infty$. Let a_n be the coordinate maps from $\mathbb{R}^{\mathbb{N}}$ into $\mathbb{R}$, then under the probability measure $\otimes_n \mu_n$, only a finite number of a_n are non zero and the scaling

$$a = \big(a_0, a_1, \dots, a_n, \dots\big) \mapsto \lambda a = \big(\lambda a_0, \lambda a_1, \dots, \lambda a_n, \dots\big) \qquad (\lambda \neq 0)$$

transforms $\otimes_n \mu_n$ into an absolutely continuous measure [equivalent measure if $\frac{d\mu}{dx} > 0$].

The proof is based upon the Borel–Cantelli lemma and an absolute continuity criterion for product measures (see Neveu [1972]) which will be left as an exercise.

Hereafter, in this Section 4, we will suppose the measure $\mathbb{P}_1 = \otimes_n \mu_n$ with μ_n chosen as above. The a_n's are the coordinate mappings of the product space.

4.3 Choice of Γ. On each factor, the choice is free provided that the structure be closable and the coordinate functions a_n be in $\mathbb{D}$. (Under $\mu_n = \alpha_n \mu + \big(1 - \alpha_n\big)\delta_0$, the Dirac mass at zero imposes the condition $\Gamma\big[a_n\big](t) = 0$ for $t = 0$.) We suppose in the sequel that the μ_n possess both first and second moments.

We will now consider two cases:

(α) $\Gamma\big[a_n\big] = a_n^2, \quad \Gamma\big[a_m, a_n\big] = 0, \quad m \neq n;$

(β) $\Gamma\big[a_n\big] = n^2 a_n^2, \quad \Gamma\big[a_m, a_n\big] = 0, \quad m \neq n.$

4.4 Analytic functions in the unit disk. Let us consider case (j). With hypotheses (α), we have

$$\Gamma[V_x](f) = \sum_n a_n^2 x^{2n}.$$

Since

$$f\left(ze^{2i\pi t}\right) = \sum_{n=0}^{\infty} a_n z^n e^{2i\pi nt},$$

using that $\left(e^{2i\pi nt}\right)_{n\in\mathbb{Z}}$ is a basis of $L^2_{\mathbb{C}}[0,1]$, we obtain

$$\Gamma[V_x](f) = \int_0^1 \left|f\left(xe^{2i\pi t}\right)\right|^2 dt \tag{21}$$

and in case (β)

$$\Gamma[V_x](f) = x^2 \int_0^1 \left|f'\left(xe^{2i\pi t}\right)\right|^2 dt. \tag{22}$$

In order to answer Question (i), let us now consider an erroneous random variable X defined on $\left(\mathbb{R}, \mathcal{B}(\mathbb{R}), \mathbb{P}_2, \mathbb{D}_2, \Gamma_2\right)$ as above and examine the error on

$$F(f, X) = f(X) = V_X(f).$$

From (20) and (21) we have in case (α)

$$\Gamma[f(X)] = \int_0^1 |f(Xe^{2i\pi t})|^2 dt + f'^2(X)\Gamma[X] \tag{23}$$

and with (22) in case (β)

$$\Gamma[f(X)] = X^2 \int_0^1 |f'(Xe^{2i\pi t})|^2 dt + f'^2(X)\Gamma[X]. \tag{24}$$

4.5 L^2-functions. The case of L^2-functions is similar. If f is represented as

$$f(x) = \sum_{n\in\mathbb{Z}} a_n e^{2i\pi nt} \quad \text{in } L^2[0,1],$$

we have

$$\Gamma[f(X)] = \Gamma[V_t](f)\big|_{t=X} + f'^2(X)\Gamma[X].$$

(Let us recall that f is almost surely a trigonometric polynomial.)

4.6 Sensitivity of an ODE. To study the sensitivity of the solution of

$$y' = f(x, y)$$

to f, let us consider the case where f is approximated by polynomials in two variables

$$f(x, y) = \sum a_{pq} x^p y^q.$$

We choose the measure $\mathbb{P}_1$ and Γ_1, as explained in 4.2 and 4.3 and in assuming measures μ_n to be centered for purpose of simplicity.

Then, if we take hypothesis (α)

$$\Gamma[a_{pq}] = a_{pq}^2$$

we obtain a sharp defined by

$$a_{pq}^{\#} = a_{pq} \frac{\widehat{a_{pq}}}{\beta_{pq}}$$

where

$$\beta_{pq} = \|a_n\|_{L^2(\mu_n)} = \left(\int_{\mathbb{R}} x^2 \, d\mu_n(x)\right)^{1/2}.$$

This sharp defines a sharp on the product space and if we consider the starting point y_0 and the value x to be random and erroneous, denoting

$$y = \varphi(x, y_0)$$

the solution to

$$\begin{cases} y' = f(x, y) \\ y(0) = y_0, \end{cases}$$

we then seek to compute $\Gamma[Y]$ for

$$Y = \varphi(X, Y_0).$$

First, suppose f alone is erroneous.

Let us remark that by the representation

$$f(t, y) = \sum_{p,q} a_{pq} t^p y^q,$$

the formula

$$(f(t, Y))^{\#} = f^{\#}(t, Y) + f_y'(t, Y) Y^{\#}$$

is still valid even when Y is not independent of f. Indeed, this expression means that

$$\left(\sum_{p,q} a_{pq} t^p Y^q\right)^{\#} = \sum_{p,q} a_{pq}^{\#} . t^p Y^q + \sum_{p,q} a_{pq} t^p q Y^{q-1} Y^{\#}$$

which is correct by the chain rule once integrability conditions have been fulfilled, thereby implying $f(t, Y) \in \mathbb{D}$.

Hence from

$$y_x = y_0 + \int_0^x f(t, y_t)\,dt,$$

we have

$$y_x^\# = \int_0^x \left(f^\#(t, y_t) + f_2'(t, y_t) y_t^\#\right) dt.$$

Let

$$M_x = \exp \int_0^x f_2'(t, y_t)\,dt$$

by the usual method of variation of the constant. This yields

$$y_x^\# = M_x \int_0^x \frac{f^\#(t, y_t)}{M_t}\,dt.$$

Now

$$f^\#(t, y_t) = \sum_{p,q} t^p y_t^p a_{pq} \frac{\widehat{a_{pq}}}{\beta_{pq}}$$

and

$$\Gamma[y_x] = M_x^2 \hat{\mathbb{E}} \left(\sum_{pq} \int_0^x \frac{t^p y_t^p a_{pq} \frac{\widehat{a_{pq}}}{\beta_{pq}}}{M_t}\,dt \right)^2.$$

Moreover

$$\Gamma[y_x] = M_x^2 \sum_{p,q} \left(\int_0^x \frac{t^p y_t^q}{M_t}\,dt \right)^2 a_{pq}^2. \tag{25}$$

If f, X and Y_0 are all three erroneous with independent settings, we obtain similarly for $Y = \varphi(X, Y_0)$

$$\Gamma[Y] = \Gamma[y_x]\Big|_{x=X, y_0=Y_0} + \varphi_1'^2(X, Y_0)\Gamma[X] + \varphi_2'^2(X, Y_0)\Gamma[Y_0],$$

however

$$\varphi_1'(x, y_0) = f(x, y_0) = \sum_{p,q} a_{pq} x^p y_0^q$$

and

$$\begin{aligned}\varphi_2'(x, y_0) = M_x &= \exp \int_0^x f_2'(t, y_t)\,dt \\ &= \exp \sum_{p,q} q a_{pq} \int_0^x t^p \varphi(t, y_0)^{q-1}\,dt.\end{aligned}$$

Thus

$$\Gamma[Y] = M_X^2 \sum_{p,q} \left(\int_0^X \frac{t^p y_t^q}{M_t} \, dt \right)^2 a_{pq}^2 \tag{26}$$
$$+ \Big(\sum_{p,q} a_{pq} X^p Y_0^q \Big)^2 \Gamma[X]$$
$$+ M_X^2 \Gamma\big[Y_0\big].$$

where $M_X = \exp\big\{ \sum_{p,q} q a_{pq} \int_0^X t^p \varphi(t, Y_0)^{q-1} \, dt \big\}$. Let us recall herein that all of these sums are finite.

Comment. The method followed in Section 4 on examples is rather general and applies in any situation where a function is approximated by a series with respect to basic functions. It reveals the sensitivity to each coefficient in the series. As seen further below, this method also applies to stochastic differential equations.

5 Substructures and projections

Let $(\Omega, \mathcal{A}, \mathbb{P}, \mathbb{D}, \Gamma)$ be an error structure. If we know the quadratic error on $Y = \big(Y_1, \dots, Y_q\big)$, what then can be said about the error on X? Is it possible to bound it from below? Is there a function of Y which represents X at its best?

Let us remark that if we can find F and $Z = \big(Z_1, \dots, Z_p\big)$ such that $X = F(Y, Z)$ and $\Gamma\big[Z_i, Y_j\big] = 0$, then we obtain the inequality

$$\Gamma[X] \geq \sum_{ij} \frac{\partial F}{\partial y_i}(Y, Z) \frac{\partial F}{\partial y_j}(Y, Z) \Gamma\big[Y_i, Y_j\big].$$

In more complex cases, we would require the notion of substructure. In this section, all error structures are assumed to be Markovian.

5.1 Error substructures.

Proposition. *Let $(\Omega, \mathcal{A}, \mathbb{P}, \mathbb{D}, \Gamma)$ be an error structure. Let V_0 be a subvector space of $\mathbb{D}$ stable by composition with $\mathcal{C}^\infty \cap \operatorname{Lip}(\mathbb{R})$ functions. Let V be the closure of V_0 in $\mathbb{D}$ and $\overline{V}$ be the closure of V (or of V_0) in $L^2(\mathbb{P})$. Then:*

1) $\overline{V} = L^2\big(\mathbb{P}|_{\sigma(V_0)}\big)$ *where $\sigma\big(V_0\big)$ is the $\mathcal{A}$-$\mathbb{P}$-complete σ-field generated by V_0.*

2) $\big(\Omega, \sigma(V_0), \mathbb{P}|_{\sigma(V_0)}, V_0, \mathbb{E}[\Gamma[\cdot] \mid \sigma(V_0)]\big)$ *is a closable error pre-structure with closure*
$$\Big(\Omega, \sigma(V_0), \mathbb{P}|_{\sigma(V_0)}, V, \mathbb{E}[\Gamma[\cdot] \mid \sigma(V_0)]\Big)$$
called the substructure generated by V_0.

As a consequence, the space V is necessarily stable by composition with Lipschitz functions of several variables.

For the proof, see Bouleau and Hirsch [1991], p. 223.

Let $\left(Y_i\right)_{i\in\mathcal{J}}$ be a family of elements of $\mathbb{D}$. Let us set

$$V_0 = \left\{G\left(Y_{i_1}, \ldots, Y_{i_k}\right), k \in \mathbb{N}, \left\{i_1, \ldots, i_k\right\} \subset \mathcal{J}, G \in \mathcal{C}^\infty \cap \mathrm{Lip}\right\},$$

then the preceding proposition applies. The space V is denoted $\mathbb{D}\left(Y_i, i \in \mathcal{J}\right)$ and called the Dirichlet sub-space generated by the family $\left(Y_i, i \in \mathcal{J}\right)$. The substructure generated by V_0 is also called the substructure generated by the family $\left(Y_i, i \in \mathcal{J}\right)$.

5.2 Remarks on projection and conditional calculus. Let $Y = \left(Y_1, \ldots, Y_q\right) \in \mathbb{D}^q$ and let us suppose the existence of a gradient D with values in $\mathcal{H}$. We then define

$$\mathrm{Grad}(Y) = \overline{\left\{ZD[U], U \in \mathbb{D}(Y), Z \in L^\infty(\mathbb{P})\right\}}^{L^2(\mathbb{P},\mathcal{H})},$$

where $\mathbb{D}(Y)$ denotes the Dirichlet subspace generated by Y, as above.

Thanks to this space, D. Nualart and M. Zakai [1988] developed a calculus by considering a conditional gradient $D[X|Y]$ to be the projection of $D[X]$ on the orthogonal of $\mathrm{Grad}(Y)$ in $L^2(\mathbb{P}, \mathcal{H})$. This notion is useful and yields nontrivial mathematical results (see Nualart–Zakai [1988], Bouleau and Hirsch [1991], Chapter V, Section 5).

Herein, we are seeking a representative of X in $\mathbb{D}(Y)$. We proceed by setting

$$D_Y[X] = \text{projection of } D[X] \text{ on } \mathrm{Grad}(Y),$$

and if X_1 and X_2 are in $\mathbb{D}$ we define

$$\Gamma_Y\left[X_1, X_2\right] = \left\langle D_Y[X_1], D_Y[X_2]\right\rangle_{\mathcal{H}}.$$

We then have:

Proposition V.10. *Suppose* $\det \Gamma\left[Y, Y^t\right] > 0$ $\mathbb{P}$*-a.e., then*

$$D_Y\left[X_1\right] = \Gamma\left[X_1, Y^t\right]\Gamma\left[Y, Y^t\right]^{-1} D[Y]$$

and for $X = \left(X_1, \ldots, X_p\right)$

$$\Gamma_Y\left[X, X^t\right] = \Gamma\left[X, Y^t\right]\Gamma\left[Y, Y^t\right]^{-1}\Gamma\left[Y, X^t\right].$$

Proof. Let us consider the case $p = q = 1$ for the sake of simplicity. Also consider

$$W = \Gamma[X, Y]\Gamma[Y]^{-1} D[Y],$$

$W \in L^2(\mathbb{P}, \mathcal{H})$ because

$$E\|W\|_{\mathcal{H}}^2 = \mathbb{E}\frac{\Gamma[X,Y]^2}{\Gamma[Y]} \leq \mathbb{E}\big[\Gamma[X]\big]$$

and if $U = G(Y)$,

$$\begin{aligned} E\langle W, ZD[U]\rangle_{\mathcal{H}} &= \mathbb{E}\left[\frac{\Gamma[X,Y]}{\Gamma[Y]} \cdot Z\langle D[Y], D[G(Y)]\rangle\right] \\ &= \mathbb{E}\left[\frac{\Gamma[X,Y]}{\Gamma[Y]} Z\Gamma[Y]G'(Y)\right] \\ &= \mathbb{E}\big[\Gamma[X, G(Y)]Z\big] \\ &= \mathbb{E}\big[\langle X, ZD[U]\rangle_{\mathcal{H}}\big]. \end{aligned}$$

Since $W \in \operatorname{Grad}(Y)$, it follows that W is the projection of X on $\operatorname{Grad}(Y)$. ◇

Exercise V.11. From this proposition, show that if $X \in \mathbb{D}$ and $Y \in \mathbb{D}$ such that $\Gamma[X] \cdot \Gamma[Y] \neq 0$ $\mathbb{P}$-a.e., then

$$\big|\langle D_X[Y], D_Y[X]\rangle\big| \leq \big|\Gamma[X,Y]\big|.$$

Remark V.12. $D_Y[X]$ is not generally $\sigma(Y)$-measurable.

Furthermore, it may be that $X \in \mathbb{D}$ and X is $\sigma(Y)$-measurable despite X not belonging to the Dirichlet subspace $\mathbb{D}(Y)$.

For example, with the structure

$$\Big([0,1], \mathcal{B}\big([0,1]\big), dx, H^1\big([0,1]\big), u \to u'^2\Big)$$

if $Y = \big(Y_1, Y_2\big) = (\sin \pi x, \sin 2\pi x)$, we have $\sigma(Y) = \mathcal{B}\big([0,1]\big)$. However, $\mathbb{D}(Y)$ is a subspace of H^1 of function u s.t. $\lim_{t\to 0} u(t) = \lim_{t\to 1} u(t)$, and there are $X \in H^1$ devoid of this property. ◇

Now consider the mixed norms

$$\|\cdot\|_\alpha = \Big(\alpha\|\cdot\|_{L^2}^2 + (1-\alpha)\mathcal{E}[\cdot]\Big)^{1/2}.$$

For $\alpha \in]0,1[$, these norms are equivalent to $\|\cdot\|_{\mathbb{D}}$ and $\mathbb{D}(Y)$ is closed in $\mathbb{D}$. We can thus consider the projection $p_\alpha(X)$ of X on $\mathbb{D}(Y)$ for the scalar product associated with $\|\cdot\|_\alpha$.

Exercise V.13. The mapping $\alpha \to p_\alpha(X)$ is continuous from $]0,1[$ into $\mathbb{D}$. When $\alpha \to 1$, $p_\alpha(X)$ converges weakly in $L^2\big(\sigma(Y)\big)$ to $\mathbb{E}[X \mid Y]$.

As $\alpha \to 0$, the question arises of the existence of a projection for $\mathcal{E}$ alone. The case $q = 1$ (Y with values in $\mathbb{R}$) can be nearly completely studied with the additional assumption of

(R) $\forall u \in \mathbb{D} \quad \mathcal{E}[u] = 0 \Rightarrow u$ is constant $\mathbb{P}$-a.e.

Lemma V.14. *Assume* (R). *If $Y \in \mathbb{D}$, the measure*

$$Y_*\big(\Gamma[Y] \cdot \mathbb{P}\big)$$

is equivalent to the Lebesgue measure on the interval (essinf Y, esssup Y).

Proof. We know that this measure is absolutely continuous.

Let F be Lipschitz. If $\Gamma\big[F(Y), Y\big] \equiv 0$, then $\mathcal{E}\big[F(Y)\big] = 0$, hence F is constant $\mathbb{P}_Y$-a.s. where $\mathbb{P}_Y$ is the law of Y.

Let B be a Borel subset $\big(Y_* \, \Gamma[Y] \cdot \mathbb{P}\big)$-negligible and let us take

$$F(y) = \int_a^y 1_B(t)\, dt.$$

It then follows from $\Gamma\big[F(Y), Y\big] = F'(Y)\Gamma[Y]$ that F is constant $\mathbb{P}_Y$-a.s.

Since the function F is continuous and vanishes at a, the property easily follows. ◇

Let us remark that the property (R) is stable by product:

Proposition V.15. *Let $S = \prod_n \big(\Omega_n, \mathcal{A}_n, \mathbb{P}_n, \mathbb{D}_n, \Gamma_n\big)$ be a product structure. If each factor satisfies* (R), *then S also satisfies* (R).

Proof. Let us denote ξ_n the coordinate mappings and let $\mathcal{F}_n = \sigma\big(\xi_m, m \le n\big)$. Then, $\mathbb{E}\big[\cdot \mid \mathcal{F}_n\big]$ is an orthogonal projector in $\mathbb{D}$ and if $F \in \mathbb{D}$, $F_n = \mathbb{E}\big[F \mid \mathcal{F}_n\big]$ converges to F in $\mathbb{D}$ as $n \uparrow \infty$.

Let $F \in \mathbb{D}$ such that $\mathcal{E}[F] = 0$, by $\mathcal{E}\big[F_n\big] \le \mathcal{E}[F]$, $\Gamma_1\big[F_n\big] + \cdots + \Gamma_n\big[F_n\big] = 0$, thus F_n does not depend on ξ_1, nor on $\xi_2, \ldots,$ nor on ξ_n and is therefore constant, hence so is F. ◇

Proposition V.16. *Suppose $S = (\Omega, \mathcal{A}, \mathbb{P}, \mathbb{D}, \Gamma)$ satisfies assumption* (R). *Let $X \in \mathbb{D}$ and $Y \in \mathbb{D}$. If*

$$\frac{\mathbb{E}\big[\Gamma[X, Y] \mid Y = y\big]}{\mathbb{E}\big[\Gamma[Y] \mid Y = y\big]}$$

is bounded $\mathbb{P}_Y$-a.s., the random variable

$$\varphi(Y) = \int_a^Y \frac{\mathbb{E}\big[\Gamma[X, Y] \mid Y = y\big]}{\mathbb{E}\big[\Gamma[Y] \mid Y = y\big]}\, dy$$

belongs to $\mathbb{D}(Y)$ and achieves the minimum of $\mathcal{E}[X - \Phi]$ among $\Phi \in \mathbb{D}(Y)$. Moreover, it is unique.

In this statement, the function of y is (according to the lemma) assumed to be zero outside of the interval (essinf Y, esssup Y).

Proof. φ is Lipschitz and satisfies

$$\varphi'(Y) = \frac{\mathbb{E}\big[\Gamma[X,Y] \mid \sigma(Y)\big]}{\mathbb{E}\big[\Gamma[Y] \mid \sigma(Y)\big]}.$$

It then follows that for every $F \in \mathcal{C}^1 \cap \mathrm{Lip}$,

$$\mathbb{E}\big[\Gamma[X,Y]F'(Y)\big] = \mathbb{E}\big[\Gamma[\varphi(Y),Y]F'(Y)\big]$$

i.e.

$$\mathcal{E}\big[X, F(Y)\big] = \mathcal{E}\big[\varphi(Y), F(Y)\big].$$

This provides the required result. ◇

The hypothesis of this proposition will be satisfied in all of the structures encountered in the following chapter.

Bibliography for Chapter V

N. Bouleau, Propriétés d'invariance du domaine du générateur étendu d'un processus de Markov, in: *Séminaire de Probabilités XV*, Lecture Notes in Math. 850, Springer-Verlag, 1981.

N. Bouleau and F. Hirsch, *Dirichlet Forms and Analysis on Wiener Space*, Walter de Gruyter, 1991.

D. Feyel, Propriétés de permanence du domaine d'un générateur infinitésimal, in: *Séminaire Théorie du Potentiel, Paris, No4*, Lecture Notes in Math. 713, Springer-Verlag, 1978.

E. Fournié, J.-M. Lasry, J. Lebuchoux, P.-L. Lions and N. Touzi, Application of Malliavin calculus to Monte Carlo methods in finance, *Finance and Stochastics* **3** (1999), 391–412.

H. Kunita, Absolute continuity of Markov processes and generators, *Nagoya J. Math.* 36 (1969), 1–26.

D. Lamberton and B. Lapeyre, *Introduction to Stochastic Calculus Applied to Finance*, Chapman & Hall, London, 1995.

P. Malliavin, *Stochastic Analysis*, Springer-Verlag, 1997.

P.-A. Meyer, Démonstration probabiliste de certaines inégalités de Littlewood-Paley, Exposé II, in: *Sém. Probab. X*, Lecture Notes in Math. 511, Springer-Verlag 1976.

G. Mokobodzki, Sur l'algèbre contenue dans le domaine étendu d'un générateur infinitésimal, *Sém. Théorie du Potentiel, Paris, No3*, Lecture Notes in Math. 681, Springer-Verlag, 1978.

J. Neveu, *Martingales à temps discret*, Masson, 1972.

D. Nualart and M. Zakai, The partial Malliavin calculus, in: *Sém. Probab. XXIII*, Lecture Notes in Math. 1372, Springer-Verlag, 1988, 362–381.

Chapter VI

Error structures on fundamental spaces space

We will now construct several error structures on the three main spaces of probability theory.

Starting with these structures it is possible, by means of images and products, to obtain error structures on the usual stochastic processes. However studying these derived structures is not compulsory, provided the usual stochastic processes have been probabilistically defined in terms of the three fundamental probability spaces, in general it suffices to consider error structures on these fundamental spaces.

1 Error structures on the Monte Carlo space

We refer to Monte Carlo space as the probability space used in simulation:

$$(\Omega, \mathcal{A}, \mathbb{P}) = \big([0,1], \mathcal{B}([0,1]), dx\big)^{\mathbb{N}}.$$

We denote the coordinate mappings $\big(U_n\big)_{n\geq 0}$. They are i.i.d. random variables uniformly distributed on the unit interval.

To obtain an error structure on this space, using the theorem on products, it suffices to choose an error structure on each factor; many (uncountably many) solutions exist.

We will begin by focus on two very simple (shift-invariant) structures useful in applications.

1.1 Structure without border terms. Consider the pre-structure

$$\Big([0,1], \mathcal{B}[0,1]\big), dx, \mathcal{C}^1([0,1]), \gamma\Big)$$

with $\gamma[u](x) = x^2(1-x)^2 u'^2(x)$ for $u \in \mathcal{C}^1[0,1]$. Let

$$\big([0,1], \mathcal{B}([0,1]), dx, \boldsymbol{d}, \gamma\big)$$

the associated error structure. We then have

Lemma VI.1.

$$\boldsymbol{d} \subset \big\{u \in L^2([0,1]) : x(1-x)u \in H_0^1(]0,1[)\big\}$$

and for $u \in \boldsymbol{d}$

$$\gamma[u] = x^2(1-x)^2 u'^2(x).$$

Proof. It is classical and straightforward to show that any u in $\boldsymbol{d}$ belongs to $H^1_{\text{loc}}\big(]0,1[\big)$ and possesses a continuous version on $]0,1[$.

Let $u \in \boldsymbol{d}$ and let $u_n \in \mathcal{C}^1[0,1]$, such that

$$u_n \to u \quad \text{in } L^2\big([0,1]\big)$$

and

$$x(1-x)u_n' \to v \quad \text{in } L^2\big([0,1]\big).$$

Let us set $h(x) = x(1-x)$ and examine the equicontinuity of the functions $h^\alpha u_n$. For $x, y \in [0,1]$ we obtain

$$\big|h^\alpha u_n(y) - h^\alpha u_n(x)\big| \le \left|\int_x^y (h^\alpha)' u_n\right| + \left|\int_x^y h^\alpha u_n'\right|$$

$$\left|\int_x^y (h^\alpha)' u_n\right| \le \left(\int_0^1 u_n^2(t)\,dt\right)^{1/2} \left(\int_x^y (h^\alpha)'^2(t)\,dt\right)^{1/2}$$

$$\left|\int_x^y h^\alpha u_n'\right| \le \left(\int_0^1 (hu_n')^2(t)\,dt\right)^{1/2} \left(\int_x^y h^{2(\alpha-1)}(t)\,dt\right)^{1/2}.$$

We observe that for $\alpha > \frac{1}{2}$, the family $h^\alpha u_n$ is equicontinuous (in both x and n). It follows then that $h^\alpha u$ possesses a continuous version with limits at 0 and 1.

Hence for $\alpha = 1$ these limits vanish. We have proved that $x(1-x)u \in H^1_0\big(]0,1[\big)$. ◇

For the generator we now have:

Lemma VI.2. *The domain* $\mathcal{D}A$ *contains* $\mathcal{C}^2\big([0,1]\big)$ *and also the functions* $u \in \mathcal{C}^2\big(]0,1[\big)$ *such that* $u \in \boldsymbol{d}$, *with* $u'x(1-x)$ *bounded and* $u''x^2(1-x)^2 \in L^2[0,1]$. *On these functions*

$$\begin{aligned} A[u](x) &= \frac{1}{2}x^2(1-x)^2 u'' + x(1-x)(1-2x)u' \\ &= u'A[I] + \frac{1}{2}u''\Gamma[I]. \end{aligned}$$

Proof. We have $u \in \mathcal{D}A$ if and only if $u \in \boldsymbol{d}$ and $v \to \mathcal{E}[u,v]$ is continuous on $\boldsymbol{d}$ for the L^2-norm. From

$$\begin{aligned} 2\mathcal{E}[u,v] &= \int_0^1 u'v'x^2(1-x)^2\,dx \\ &= \big[u'vx^2(1-x)^2\big]_0^1 - \int_0^1 \big(u'x^2(1-x)^2\big)' v\,dx \end{aligned}$$

we easily obtain the result with the reliance on the preceding proof. $\diamond$

Let us now consider the product structure

$$(\Omega, \mathcal{A}, \mathbb{P}, \mathbb{D}, \Gamma) = \big([0,1], \mathcal{B}([0,1]), dx, \boldsymbol{d}, \gamma\big)^{\mathbb{N}}$$

The coordinate mappings U_n verify

$$\begin{aligned} U_n &\in \mathcal{D}A \subset \mathbb{D} \\ \Gamma[U_n] &= U_n^2(1-U_n)^2 \\ \Gamma[U_m, U_n] &= 0 \quad \forall m \neq n \\ A[U_n] &= U_n(1-U_n)(1-2U_n). \end{aligned}$$

Set $\varphi(x) = x^2(1-x)^2$.

1.1.1. According to the theorem on products, if $F = f(U_0, U_1, \ldots, U_n, \ldots)$ is a real random variable, then $F \in \mathbb{D}$ if and only if

$$\forall n \;\; x \to f(U_0, \ldots, U_{n-1}, x, U_{n+1}, \ldots) \in \boldsymbol{d}$$

and

$$\mathbb{E}\Big[\sum_n f_n'^2(U_0, U_1, \ldots)\varphi(U_n)\Big] < +\infty.$$

1.1.2. We can define a gradient with $\mathcal{H} = \ell^2$ and

$$DF = \Big(f_n'(U_0, U_1, \ldots)\sqrt{\varphi(U_n)}\Big)_{n\in\mathbb{N}}.$$

1.1.3. If $a \in \ell^2$, $a = (a_n)$, we can easily observe that $a \in \operatorname{dom} \delta$ and $\delta[a] = \sum_n a_n(2U_n - 1)$ (which is a square integrable martingale) such that, $\forall F \in \mathbb{D}$

$$\mathbb{E}[\langle DF, a\rangle] = \mathbb{E}\Big[F \sum_n a_n(2U_n - 1)\Big].$$

1.1.4. Applying this relation to FG for $F, G \in \mathbb{D} \cap L^\infty$ yields

$$\mathbb{E}[G\langle DF, a\rangle_{\ell^2}] = \mathbb{E}\Big[F\Big(G\sum_n a_n(2U_n - 1) - \langle DG, a\rangle\Big)\Big].$$

Proposition VI.3. *Let $Y = (Y_n(U_0, U_1, \ldots))_{n\geq 0}$ be a sequence of bounded random variables Y_n such that $\mathbb{E}\sum_n Y_n^2 < +\infty$, Y_n is $\mathcal{C}^1$ with respect to U_n and the series*

$$\sum_n \frac{\partial}{\partial U_n}\big(U_n(1-U_n)Y_n\big)$$

converges in $L^2(\mathbb{P})$, then $Y \in \operatorname{dom}\delta$ and

$$\delta[Y] = -\sum_n \frac{\partial}{\partial U_n}\big(U_n(1-U_n)Y_n\big).$$

Proof. With these hypotheses, $Y \in L^2(\mathbb{P}, \mathcal{H})$ and $\forall F \in \mathbb{D}$, $F = f\big(U_0, U_1, \dots\big)$

$$\mathbb{E}\langle DF, Y\rangle = \mathbb{E}\sum_n f'_n \sqrt{\varphi\big(U_n\big)}\, Y_n$$

and thanks to Lemma VI.1 and the above observation 1.1.1, this also yields

$$= -\mathbb{E}\sum_n F \frac{\partial}{\partial U_n}\Big(\sqrt{\varphi(U_n)}\, Y_n\Big)$$

which provides the result by the definition of δ and dom δ. ◇

1.1.5. If $F = f\big(U_0, U_1, \dots\big)$ is of class $\mathcal{C}^2\big([0,1]\big)$ in each variable and if the series

$$\sum_n f'_n A\big[U_n\big] + \frac{1}{2} f''_{nn} \Gamma\big[U_n\big]$$

converges in $L^2(\mathbb{P})$ then $F \in \mathcal{D}A$ and

$$A[F] = \sum_n f'_n \cdot A\big[U_n\big] + \frac{1}{2} f''_{nn} \cdot \Gamma\big[U_n\big].$$

1.2 Example (Sensitivity of a simulated Markov chain). Suppose by discretization of an SDE with respect to a martingale or a process with independent increments, we obtain a Markov chain

$$\begin{cases} S_0 = x \\ S_{n+1} = S_n + \sigma\big(S_n\big)\big(Y_{n+1} - Y_n\big) \end{cases}$$

where Y_n is a martingale simulated by

$$Y_n - Y_{n-1} = \xi\big(n, U_n\big) \quad \text{with} \int_0^1 \xi(n,x)\,dx = 0$$

and where σ is a Lipschitz function.

1.2.1 Sensitivity to the starting point. Here, we are not placing any error on the U_n's but we suppose that $S_0 = x$ is erroneous.

For example, let us take on the starting point the Ornstein–Uhlenbeck structure

$$\big(\mathbb{R}, \mathcal{B}(\mathbb{R}), m, H^1(m), u \to u'^2\big)$$

with $m = \mathcal{N}(0, 1)$.

The computation is obvious; the sole advantage provided by error structures is to be able to work with Lipschitz hypotheses only.

We obtain for Ψ Lipschitz

$$\Gamma[\Psi(S_N)] = \Big(\Psi'(S_N) \prod_{i=1}^{N-1} \big(1 + \sigma'(S_i)(Y_{i+1} - Y_i)\big)\Big)^2 \big(1 + \sigma'(x)(Y_1 - Y_0)\big)^2$$

and for σ and Ψ $\mathcal{C}^2$:

$$\begin{aligned} A[\Psi(S_N)] &= \Psi'(S_N)A[S_N] + \frac{1}{2}\Psi''(S_N)\Gamma[S_N] \\ A[S_N] &= A[S_{N-1}] + \sigma'(S_{N-1})A[S_{N-1}](Y_N - Y_{N-1}) \\ &\quad + \frac{1}{2}\sigma''(S_{N-1})\Gamma[S_{N-1}](Y_N - Y_{N-1}). \end{aligned}$$

We can remark that $A[S_N]$ is a martingale and $\frac{\Gamma[\Psi(S_N)]}{\Psi'^2(S_N)}$ a sub-martingale.

1.2.2 Internalization for $\frac{d}{dx}\mathbb{E}[\Psi(S_N)]$. We are seeking $\psi(\omega, x)$ such that

$$\frac{d}{dx}\mathbb{E}[\Psi(S_N)] = \mathbb{E}[\Psi(S_N)\psi(\omega, x)].$$

We place an error only on U_1 and choose the error structure of Lemma VI.1. Assuming $\xi(1, x)$ to be $\mathcal{C}^2$ in x and using the integration by parts formula

$$\mathbb{E}[GDF] = -\mathbb{E}\left[F\frac{d}{dU_1}(G\sqrt{\varphi(U_1)})\right]$$

leads to

$$\frac{d}{dx}\mathbb{E}[\Psi(S_N)] = \mathbb{E}\left[\Psi(S_N)\left(\frac{\xi''_{x^2}(1, U_1)(1 + \sigma'(x)\xi(1, U_1))}{\sigma(x)\xi'^2_x(1, U_1)} - \frac{\sigma'(x)}{\sigma(x)}\right)\right].$$

1.2.3 Deriving information on the law of S_N. We introduce an error on each U_n and then work with the Monte Carlo space using the above defined structure. If $F = f(U_1, \dots, U_n, \dots)$ we obtain

$$DF = \Big(f'_i\sqrt{\varphi(U_i)}\Big)_{i\geq 1}.$$

With $F = \Psi(S_N)$, we get $DF = \Psi'(S_N)DS_N$, i.e.,

$$\begin{aligned} f'_N\sqrt{\varphi(U_N)} &= \Psi'(S_N)\sigma(S_{N-1})\xi'(N,U_N)\sqrt{\varphi(U_N)} \\ f'_{N-1}\sqrt{\varphi(U_{N-1})} &= \Psi'(S_N)(1+\sigma'(S_{N-1})\xi(N,U_N)) \\ &\quad\times \sigma(S_{N-2})\xi'(N-1,U_{N-1})\sqrt{\varphi(U_{N-1})} \\ &\ \ \vdots \\ f'_1\sqrt{\varphi(U_1)} &= \Psi'(S_N)\Big(\prod_{k=1}^{N}(1+\sigma'(S_{k-1})\xi(k,U_k))\Big)\sigma(x)\xi'(1,U_1)\sqrt{\varphi(U_1)}. \end{aligned}$$

For $a \in \ell^2$, using the IPF (integration by parts formula)

$$\mathbb{E}\big[G\langle DF, a\rangle\big] = -\mathbb{E}\Big[F\Big(\langle DG, a\rangle - G\sum_n a_n(2U_n-1)\Big)\Big]$$

with $G = \frac{1}{\langle DS_N, a\rangle}$ yields the relation

$$\mathbb{E}\big[\Psi'(S_N)\big] = \mathbb{E}\left[\Psi(S_N)\left(\frac{\sum_n a_n(2U_n-1)}{\langle DS_N, a\rangle} - \left\langle D\frac{1}{\langle DS_N, a\rangle}, a\right\rangle\right)\right]$$

which can, with a suitable assumption, yield the regularity of the law of S_N, this is Malliavin's method.

According to the density criterion (Theorem III.12 of Chapter III) we now have with the Lipschitz hypotheses

$$\begin{aligned} \Gamma\big[S_N\big] &= \sigma^2(S_{N-1})\xi'^2(N,U_N)\varphi(U_N) \\ &\quad + \cdots + \Big(\prod_{k=1}^{N}(1+\sigma'(S_{k-1})\xi(k,U_k))\Big)^2\big(\sigma(x)\xi'(1,U_1)\big)^2\varphi(U_1). \end{aligned}$$

We observe that if $\sigma(x) \neq 0$ and $\xi'(k,x) \neq 0\ \forall k$, then S_N has density.

Moreover, observing from the above calculation of DS_N that

$$\det\begin{pmatrix} \Gamma[S_N] & \Gamma[S_N, S_{N-1}] \\ \Gamma[S_N, S_{N-1}] & \Gamma[S_{N-1}] \end{pmatrix} = \Gamma\big[S_{N-1}\big]\big(\sigma^2(S_{N-1})\xi'^2\varphi(U_N)\big) > 0,$$

we obtain from Propositions III.16 and IV.10 that the pair (S_N, S_{N-1}) also has a density.

1.3 The pseudo-Gaussian structure on the Monte Carlo space. Consider the Ornstein–Uhlenbeck structure on $\mathbb{R}$,

$$\left(\mathbb{R}, \mathcal{B}(\mathbb{R}), m, H^1(m), u \to u'^2\right)$$

with $m = \mathcal{N}(0,1)$.

Let us denote $N(x) = \int_{-\infty}^{x} \frac{1}{\sqrt{2\pi}} e^{-\frac{t^2}{2}}\, dt$ the distribution function of the reduced normal law and

$$\varphi_1(x) = \frac{1}{2\pi} \exp -\left(N^{-1}(x)\right)^2,$$

the image by N of the Ornstein–Uhlenbeck structure then gives a structure on $\big([0,1], \mathcal{B}\big([0,1]\big), dx\big)$, i.e.

$$\big([0,1], \mathcal{B}\big([0,1]\big), dx, \boldsymbol{d}_1, \gamma_1\big)$$

with

$$\begin{aligned} \boldsymbol{d}_1 &= \left\{u \in L^2[0,1] : u \circ N \in H^1(m)\right\} \\ \gamma_1[u](x) &= \varphi_1(x) u'^2(x). \end{aligned}$$

Although the function φ_1 is not as simple as φ, this structure still possesses an IPF without border terms like the preceding structure, and gives rise to an efficient IPF on the Monte Carlo space.

Another interesting property of this structure is that it satisfies a Poincaré-type inequality.

Proposition VI.4. *Let S be the product*

$$S = \big([0,1], \mathcal{B}\big([0,1]\big), dx, \boldsymbol{d}_1, \gamma_1\big)^{\mathbb{N}}.$$

Then $\forall F \in \mathbb{D}$ we the following inequalities hold:

$$\operatorname{var}[F] = \mathbb{E}\big[(F - \mathbb{E}F)^2\big] \le 2\mathcal{E}[F].$$

Proof. This property is true for the Ornstein–Uhlenbeck structure (see D. Chafaï [2000]) and is preserved by both images and products. ◇

Exercise VI.5 (Probabilistic interpretation of Γ). Consider the finite dimensional Monte Carlo space with the structure

$$\big([0,1], \mathcal{B}\big([0,1]\big), dx, H^1\big([0,1]\big), u \to u'^2\big)^k.$$

Let P_t be the transition semigroup of the Brownian motion in the cube $[0,1]^k$ with reflection on the boundary. Let $F: [0,1]^k \to \mathbb{R}$ be Lipschitz, then

$$\lim_{t \downarrow 0} \frac{1}{t} P_t\big(F - F(x)\big)^2(x) = \Gamma[F](x)$$

for dx-almost every $x \in [0,1]^k$, and the limit remains bounded by K^2, where K is the Lipschitz constant of F.

In order to better understand this set-up let us introduce the mapping from $\mathbb{R}^k$ onto $[0,1]^k$ induced by reflections on the boundary

$$x \in \mathbb{R}^k \xrightarrow{\eta} y = \eta(x) \in [0,1]^k.$$

When $k = 2$, this mapping can be described by folding a paper plane alternatively up and down along the straight lines $\dots, x = -1, x = 0, x = 1, \dots$ and then alternatively up and down along the lines $\dots y = -1, y = 0, y = 1, \dots$ so that a square is ultimately obtained. In more formal terms, y is the unique point in $[0,1]^k$ such that

$$x \in \big\{(2n_1 \pm y_1, \dots, 2n_k \pm y_k) \text{ with } n_1, \dots n_k \in \mathbb{Z}\big\}.$$

Then, let us set $\tilde{F}(x) = F\big(\eta(x)\big)$ and denote $\mathbb{P}_0$ the law of the standard Brownian motion starting from 0 in $\mathbb{R}^k$, and $\mathbb{E}_0$ the corresponding expectation.

For $x \in [0,1]^k$, we obtain

$$\begin{aligned}\frac{1}{t} P_t\big(F - F(x)\big)^2(x) &= \mathbb{E}_0 \frac{\big(\tilde{F}(x + B_t) - \tilde{F}(x)\big)^2}{t} \\ &= \mathbb{E}_0 \frac{\big(\tilde{F}(x + \sqrt{t} B_1) - \tilde{F}(x)\big)^2}{t}.\end{aligned}$$

The function $\tilde{F}$ is Lipschitz with the same constant as F and a.e. differentiable:

For dx-a.e. x we have

$$\forall \omega \ \lim_{t \downarrow 0} \frac{\tilde{F}\big(x + \sqrt{t} B_1(\omega)\big) - \tilde{F}(\omega)}{t} = B_1(\omega) \cdot \nabla \tilde{F}(x)$$

and the quotient is bounded in modulus by $K|B_1|(\omega)$, so that

$$\frac{\tilde{F}\big(x + \sqrt{t} B_1\big) - \tilde{F}(x)}{\sqrt{t}} \to B_1 \cdot \nabla \tilde{F}(x)$$

in $L^2(\mathbb{P}_0)$. For dx-a.e. x

$$\frac{1}{t} P_t\big(F - F(x)\big)^2(x) \to \mathbb{E}_0 \big|B_1 \cdot \nabla \tilde{F}(x)\big|^2 = |\nabla \tilde{F}(x)|^2.$$

The required property is thus proved. ◇

2 Error structures on the Wiener space

Let us first recall the classical approach of the so-called Wiener integral.

2.1 The Wiener stochastic integral. Let $(T, \mathcal{T}, \mu)$ be a σ-finite measured space, $(\chi_n)_{n\in\mathbb{N}}$ an orthonormal basis of $L^2(T, \mathcal{T}, \mu)$, and $(g_n)_{n\in\mathbb{N}}$ a sequence of i.i.d. reduced Gaussian variables defined on $(\Omega, \mathcal{A}, \mathbb{P})$.

If with $f \in L^2(T, \mathcal{T}, \mu)$ we associate $I(f) \in L^2(\Omega, \mathcal{A}, \mathbb{P})$ defined via

$$I(f) = \sum_n \langle f, \chi_n \rangle g_n,$$

I becomes a homomorphism from $L^2(T, \mu)$ into $L^2(\Omega, \mathcal{A}, \mathbb{P})$.

If $f, g \in L^2(T, \mathcal{T}, \mu)$ are such that $\langle f, g\rangle = 0$, then $I(f)$ and $I(g)$ are two independent Gaussian variables.

From now on, we will take either $(T, \mathcal{T}, \mu) = \big(\mathbb{R}_+, \mathcal{B}(\mathbb{R}_+), dx\big)$ or $\big([0,1], \mathcal{B}([0,1]), dx\big)$.

If we set

$$B(t) = \sum_n \big\langle 1_{[0,t]}, \chi_n \big\rangle g_n = \sum_n \int_0^t \chi_n(y)\,dy \cdot g_n \tag{1}$$

then $B(t)$ is a centered Gaussian process with covariance

$$\mathbb{E}\big[B(t)B(s)\big] = t \wedge s$$

i.e., a standard Brownian motion.

It can be shown that the series (1) converges in both $\mathcal{C}_K\big(\mathbb{R}_+\big)$ a.s. and $L^p\big((\Omega, \mathcal{A}, \mathbb{P}), \mathcal{C}_K\big)$ for $p \in [1, \infty[$ (where K denotes a compact set in $\mathbb{R}_+$).

Due to the case where f is a step-function, the random variable $I(f)$ is denoted

$$I(f) = \int_0^\infty f(s)\,dB_s \quad \text{(resp. } \int_0^1 f(s)\,dB_s)$$

and called the Wiener integral of f.

2.2 Product error structures. The preceding construction actually involves the product probability space

$$(\Omega, \mathcal{A}, \mathbb{P}) = \big(\mathbb{R}, \mathcal{B}(\mathbb{R}), \mathcal{N}(0,1)\big)^{\mathbb{N}},$$

with the g_n's being the coordinate mappings. If we place on each factor an error structure

$$\big(\mathbb{R}, \mathcal{B}(\mathbb{R}), \mathcal{N}(0,1), \boldsymbol{d}_n, \gamma_n\big),$$

we obtain an error structure on $(\Omega, \mathcal{A}, \mathbb{P})$ as follows:

$$(\Omega, \mathcal{A}, \mathbb{P}, \mathbb{D}, \Gamma) = \prod_{n=0}^{\infty} \big(\mathbb{R}, \mathcal{B}(\mathbb{R}), \mathcal{N}(0,1), \boldsymbol{d}_n, \gamma_n\big)$$

such that a random variable

$$F\big(g_0, g_1, \dots, g_n, \dots\big)$$

belongs to $\mathbb{D}$ if and only if $\forall n, x \to F\big(g_0, \dots, g_{n-1}, x, g_n, \dots\big)$ belongs to $\boldsymbol{d}_n$ $\mathbb{P}$-a.s. and

$$\Gamma[F] = \sum_n \gamma_n[F],$$

γ_n acting on the n-th variable of F, belongs to $L^1(\mathbb{P})$.

2.3 The Ornstein–Uhlenbeck structure. On each factor, we consider the one-dimensional Ornstein–Uhlenbeck structure (see Chapters II and III Example 1). Hence, we obtain

$$\Gamma\big[g_n\big] = 1$$
$$\Gamma\big[g_m, g_n\big] = 0 \quad \text{if } m \neq n.$$

For $f \in L^2(\mathbb{R}_+)$, by $\int_0^\infty f(s)\,dB_s = \sum_n \langle f, \chi_n\rangle g_n$ we obtain

$$\Gamma\left[\int_0^\infty f(s)\,dB_s\right] = \sum_n \langle f, \chi_n\rangle^2 = \|f\|^2_{L^2(\mathbb{R}_+)},$$

From the relation

$$\Gamma\left[\int_0^\infty f(s)\,dB_s\right] = \|f\|^2_{L^2(\mathbb{R}_+)} \tag{2}$$

we derive that, $\forall F \in \mathcal{C}^1 \cap \operatorname{Lip}(\mathbb{R}^m)$,

$$\Gamma\left[F\left(\int f_1(s)\,dB_s, \dots, \int f_n(s)\,dB_s\right)\right] = \sum_{i,j} \frac{\partial F}{\partial x_i}\frac{\partial F}{\partial x_j} \int f_i(s) f_j(s)\,ds.$$

This relation defines Γ on a dense subspace of $L^2(\mathbb{P})$ because it contains the $\mathcal{C}^1 \cap \operatorname{Lip}$ functions of a finite number of g_n's, which prove to be dense by virtue of the construction of the product measure. In other words, any error structure on the Wiener space such that $\mathbb{D}$ contains $\int f\,dB$ for $f \in \mathcal{C}_K^\infty(\mathbb{R}_+)$ and satisfies (2) is an extension of the Ornstein–Uhlenbeck structure, in fact coincides with it: it can be proved that (2) characterizes the Ornstein–Uhlenbeck structure on the Wiener space among the structures such that $\mathbb{D}$ contains $\int f\,dB$ for $f \in \mathcal{C}_K^\infty(\mathbb{R}_+)$.

Gradient. We can easily define a gradient operator with $\mathcal{H} = L^2(\mathbb{R}_+)$: for $G \in \mathbb{D}$ let us set

$$D[G] = \sum_n \frac{\partial G}{\partial g_n} \cdot \chi_n(t). \tag{3}$$

This approach makes sense according to the theorem on products and satisfies

$$\langle D[G], D[G] \rangle = \sum_n \left(\frac{\partial G}{\partial g_n} \right)^2 = \Gamma[G],$$

therefore D is a gradient.

For $h \in L^2(\mathbb{R}_+)$, we obtain

$$D\left[\int_0^\infty h(s)\, dB_s \right] = h \tag{4}$$

(since $\int_0^\infty h(s)\, dB_s = \sum_n \langle h, \chi_n \rangle g_n$ and $D[g_n] = \chi_n$).

Proposition VI.6. *If $h \in L^2(\mathbb{R}_+)$ and $F \in \mathbb{D}$,*

$$\mathbb{E}\langle DF, h \rangle_{\mathcal{H}} = \mathbb{E}\left[F \int_0^\infty h\, dB \right].$$

Proof. Let us adopt the notation

$$F = F\big(g_0, g_1, \ldots, g_n, \ldots\big).$$

Then

$$D[F] = \sum_n \frac{\partial F}{\partial g_n} \chi_n$$

and

$$\langle D[F], h \rangle_{\mathcal{H}} = \sum_n \frac{\partial F}{\partial g_n} \langle \chi_n, h \rangle$$

$$\mathbb{E}\big[\langle DF, h \rangle\big] = \sum_n \mathbb{E}\left[\frac{\partial F}{\partial g_n}\big(g_0, \ldots, g_n, \ldots\big) \right] \langle \chi_n, h \rangle.$$

However,

$$\begin{aligned}
\mathbb{E}\left[\frac{\partial F}{\partial g_n}\big(g_0, \ldots, g_n, \ldots\big) \right] &= \mathbb{E} \int \frac{\partial F}{\partial g_n}\big(g_0, \ldots, g_{n-1}, x, g_n, \ldots\big) \frac{1}{\sqrt{2\pi}} e^{-\frac{x^2}{2}}\, dx \\
&= \mathbb{E} \int F\big(g_0, \ldots, g_{n-1}, x, g_n, \ldots\big) x \frac{1}{\sqrt{2\pi}} e^{-\frac{x^2}{2}}\, dx \\
&= \mathbb{E}\big[g_n F(g_0, \ldots, g_n, \ldots) \big].
\end{aligned}$$

Hence

$$\mathbb{E}\big[\langle DF,h\rangle\big]=\mathbb{E}\left[F\sum_n g_n\langle\chi_n,h\rangle\right]=\mathbb{E}\left[F\int_0^\infty h\,dB\right].\qquad\diamond$$

Corollary VI.7. $\forall F,G\in\mathbb{D}\cap L^\infty$

$$\mathbb{E}\big[G\langle DF,h\rangle_{\mathcal{H}}\big]=-\mathbb{E}\big[F\langle DG,h\rangle\big]+\mathbb{E}\left[FG\int h\,dB\right].$$

Let $\mathcal{F}_t=\sigma\big(B_s,s\le t\big)$ be the natural filtration of the Brownian motion, we have

Lemma VI.8. *The operators $\mathbb{E}\big[\cdot\mid\mathcal{F}_s\big]$ are orthogonal projectors in $\mathbb{D}$, and for $X\in\mathbb{D}$*

$$D\big[\mathbb{E}[X\mid\mathcal{F}_s]\big]=\mathbb{E}\big[(DX)(t)1_{t\le s}\mid\mathcal{F}_s\big].$$

We often write D_tX for $DX(t)$.

Proof. a) It is sufficient to prove the lemma for

$$X=F\left(\int h_1\,dB,\dots,\int h_k\,dB\right)$$

with $h_i\in L^2(\mathbb{R}_+)$ and $F\in\mathcal{C}^1\cap\mathrm{Lip}$.

It is an easy exercise to demonstrate that

$$\mathbb{E}\big[X\mid\mathcal{F}_s\big]=\hat{\mathbb{E}}\left[F\left(\int_0^s h_1\,dB+\int_s^\infty h_1\,d\hat{B},\dots,\int_0^s h_k\,dB+\int_s^\infty h_k\,d\hat{B}\right)\right],$$

where $\hat{B}$ is a copy of B and $\hat{E}$ the corresponding expectation.

For the sake of simplicity, let us set

$$\begin{aligned}U&=\left(\int_0^\infty h_1\,dB,\dots,\int_0^\infty h_k\,dB\right)\\ V&=\left(\int_0^s h_1\,dB+\int_s^\infty h_1\,d\hat{B},\dots,\int_0^s h_k\,dB+\int_s^\infty h_k\,d\hat{B}\right).\end{aligned}$$

We then have

$$\begin{aligned}D\big[\mathbb{E}[X\mid\mathcal{F}_s]\big]&=\sum_i\hat{\mathbb{E}}\left[\frac{\partial F}{\partial x_i}(V)\right]h_i(t)1_{t\le s}\\ DX-D\big[\mathbb{E}[X\mid\mathcal{F}_s]\big]&=\sum_i\left(\frac{\partial F}{\partial x_i}(U)-\hat{\mathbb{E}}\left[\frac{\partial F}{\partial x_i}(V)\right]\right)h_i(t)1_{t\le s}\\ &\quad+\sum_i\frac{\partial F}{\partial x_i}(U)h_i(t)1_{t>s}\end{aligned}$$

and

$$\left\langle D\big[\mathbb{E}[X \mid \mathcal{F}_s]\big], DX - D\big[\mathbb{E}[X \mid \mathcal{F}_s]\big]\right\rangle_{\mathcal{H}}$$
$$= \sum_{i,j} \hat{\mathbb{E}}\left[\frac{\partial F}{\partial x_i}(V)\right]\left(\frac{\partial F}{\partial x_j}(U) - \hat{\mathbb{E}}\left[\frac{\partial F}{\partial x_j}(V)\right]\right) \int_0^s h_i(t)h_j(t)\,dt,$$

yet the expectation of this expression is zero, as seen by writing it $\mathbb{E}\big[\mathbb{E}[\cdot \mid \mathcal{F}_s]\big]$. Hence

$$\mathcal{E}\big[\mathbb{E}[X \mid \mathcal{F}_s], X - \mathbb{E}[X \mid \mathcal{F}_s]\big] = 0.$$

The orthogonality of the conditional expectation in $\mathbb{D}$ follows by density.

b) The same approach yields the formula in the statement. ◇

We are now able to study the adjoint operator of the gradient, the operator δ.

Proposition VI.9. *Let $u_t \in L^2(\mathbb{R}_+ \times \Omega, dt \times d\mathbb{P})$ be an adapted process (u_t is $\mathcal{F}_t$-measurable up to $\mathbb{P}$-negligible sets, $\forall t$), then $u_t \in$ dom δ and*

$$\delta[u_t] = \int_0^\infty u_t\,dB_t.$$

Thus δ extends the Itô stochastic integral and coincides with it on adapted processes.

Proof. a) Consider an elementary adapted process

$$v_t = \sum_{i=1}^n F_i 1_{]t_i,t_{i+1}]}(t)$$

where the F_i's are F_{t_i}-measurable and in L^2, these processes are dense among the adapted processes in $L^2(\mathbb{R}_+ \times \Omega, dt \times d\mathbb{P})$. We can even suppose $F_i \in \mathbb{D} \cap L^\infty$.

For $F \in \mathbb{D} \cap L^\infty$, let us study

$$\mathbb{E}\int_0^\infty DF(t) \cdot v_t\,dt.$$

From the IPF of Corollary VI.7, we can derive

$$\mathbb{E}\left[\int_0^\infty DF(t)F_i 1_{]t_i,t_{i+1}]}(t)\,dt\right] = -\mathbb{E}\big[F\langle DF_i, 1_{]t_i,t_{i+1}]}\rangle\big] + \mathbb{E}\big[FF_i\big(B_{t_{i+1}} - B_{t_i}\big)\big].$$

Hence

$$\left|\mathbb{E}\left[\int_0^\infty DF\,F_i 1_{]t_i,t_{i+1}]}\,dt\right]\right| \le c\|F\|_{L^2(\mathbb{P})}$$

and

$$\delta\big[F_i 1_{]t_i,t_{i+1}]}\big] = -\int_{t_i}^{t_{i+1}} DF_i(t)\,dt + F_i\big(B_{t_{i+1}} - B_{t_i}\big).$$

According to the preceding lemma however, the first term in the right-hand side is zero. Therefore

$$\delta[v_t] = \sum_i F_i\big(B_{t_{i+1}} - B_{t_i}\big) = \int_0^\infty v_t\,dB_t.$$

b) Since δ is a closed operator, taking a sequence of elementary adapted processes $v_n \to v$ in $L^2\big(\mathbb{R}_+ \times \Omega, dt \times d\mathbb{P}\big)$ yields

$$\int_0^\infty v_n(t)\,dB_t = \delta[v_n] \xrightarrow[n\uparrow\infty]{} \int_0^\infty u_t\,dB_t = \delta[v]. \qquad \diamond$$

The sharp. The general definition lends the following relations:

$$\big(F(g_0,\dots,g_n,\dots)\big)^\# = \sum_n \frac{\partial F}{\partial g_n}(g_0,\dots,g_n,\dots)\hat{g}_n$$

$$\forall X \in \mathbb{D} \quad X^\#(\omega,\hat{\omega}) = \int_0^\infty DX(t)\,d\hat{B}_t$$

$$\Gamma[X] = \hat{\mathbb{E}}\big[X^{\#2}\big].$$

From (4) we obtain

$$\left(\int_0^\infty h(s)\,dB_s\right)^\# = \int_0^\infty h(s)\,d\hat{B}_s.$$

Proposition VI.10. *Let u be an adapted process in the closure of the space*

$$\left\{\sum_{i=1}^n F_i 1_{]t_i,t_{i+1}]},\ F_i \in \mathcal{F}_{t_i},\ F_i \in \mathbb{D}\right\}$$

for the norm $\big(\mathbb{E}\int_0^\infty u^2(s)ds + \mathbb{E}\int_0^\infty\int_0^\infty (D_t[u(s)])^2\,dsdt\big)^{1/2}$. *Then*

$$\left(\int_0^\infty u_s\,dB_s\right)^\# = \int_0^\infty (u_s)^\#\,dB_s + \int_0^\infty u_s\,d\hat{B}_s.$$

The proof proceeds by approximation and is left to the reader. $\diamond$

As an application of the sharp, we propose the following exercises.

Exercise VI.11. Let $f_1(s,t)$ and $f_2(s,t)$ belong to $L^2\big(\mathbb{R}_+^2, ds\,dt\big)$ and be symmetric.

Let $U = \big(U_1, U_2\big)$ with

$$U_i = \int_0^\infty \int_0^t f_i(s,t)\,dB_s\,dB_t, \quad i = 1, 2.$$

If $\det\big(\Gamma[U, U^t]\big) = 0$ a.e. then the law of U is carried by a straight line.

Hint. Show that

$$U_i^\# = \int_0^\infty \int_0^\infty f_i(s,t)\,dB_s\,d\hat{B}_t.$$

From $\big(\hat{\mathbb{E}}[U_1^\# U_2^\#]\big)^2 = \hat{\mathbb{E}}[U_1^{\#2}]\hat{\mathbb{E}}[U_2^{\#2}]$ deduce that a random variable $A(\omega)$ exists whereby

$$U_1^\#(\omega, \hat{\omega}) = A(\omega) U_2^\#(\omega, \hat{\omega}).$$

Use the symmetry of $U_1^\#$ and $U_2^\#$ in $(\omega, \hat{\omega})$ in order to deduce that A is constant. ◇

Exercise VI.12. Let $f(s,t)$ be as in the preceding exercise, and g belong to $L^2(\mathbb{R}_+)$. If

$$X = \int_0^\infty g(s)\,dB_s$$
$$Y = \int_0^\infty \int_0^t f(s,t)\,dB_s\,dB_t$$

show that

$$\Gamma[X] = \|g\|_{L^2}^2$$
$$\Gamma[Y] = \int_0^\infty \left(\int_0^\infty f(s,t)\,dB_s\right)^2 dt$$
$$\Gamma[X, Y] = \int_0^\infty g(s) \left(\int_0^\infty f(s,t)\,dB_t\right) ds.$$

Show that if $\big(\Gamma[X,Y]\big)^2 = \Gamma[X]\Gamma[Y]$, the law of (X, Y) is carried by a parabola.

Numerical application. Let us consider the case

$$f(s,t) = 2h(s)h(t) - 2g(s)g(t)$$

for $g, h \in L^2(\mathbb{R}_+)$ with $\|h\|_{L^2} = \|g\|_{L^2} = 1$ and $\langle g, h\rangle = 0$.

The pair (X, Y) then possesses the density

$$\frac{1}{4\pi} e^{-y/2} e^{-x^2} \frac{1}{\sqrt{y + x^2}} 1_{\{y > -x^2\}},$$

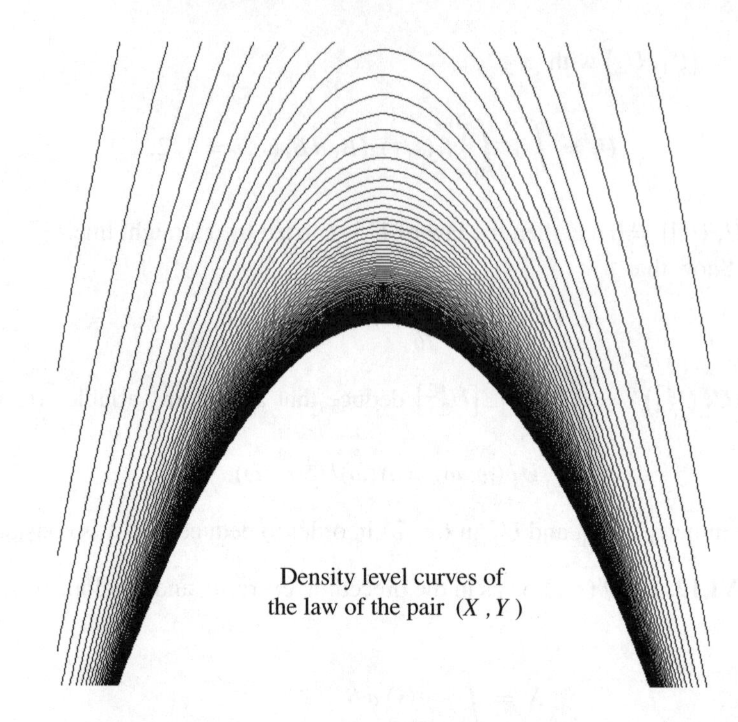

and the matrix of the error variances is

$$\begin{pmatrix} \Gamma[X] & \Gamma[X,Y] \\ \Gamma[X,Y] & \Gamma[Y] \end{pmatrix} = \begin{pmatrix} 1 & -2X \\ -2X & 8X^2 + 4Y \end{pmatrix}.$$

In other words, the image error structure by (X, Y) possesses a quadratic error operator $\Gamma_{(X,Y)}$ such that for $\mathcal{C}^1 \cap$ Lip-functions

$$\Gamma_{(X,Y)}[F](x,y) = F'^2_1(x,y) - 4xF'_1(x,y)F'_2(x,y) + (8x^2 + 4y)F'^2_2(x,y).$$

This can be graphically represented, as explained in Chapter I, by a field of ellipses of equations

$$(u\ v) \begin{pmatrix} 1 & -2x \\ -2x & 8x^2 + 4y \end{pmatrix}^{-1} \begin{pmatrix} u \\ v \end{pmatrix} = \varepsilon^2$$

which are the level curves of small Gaussian densities.

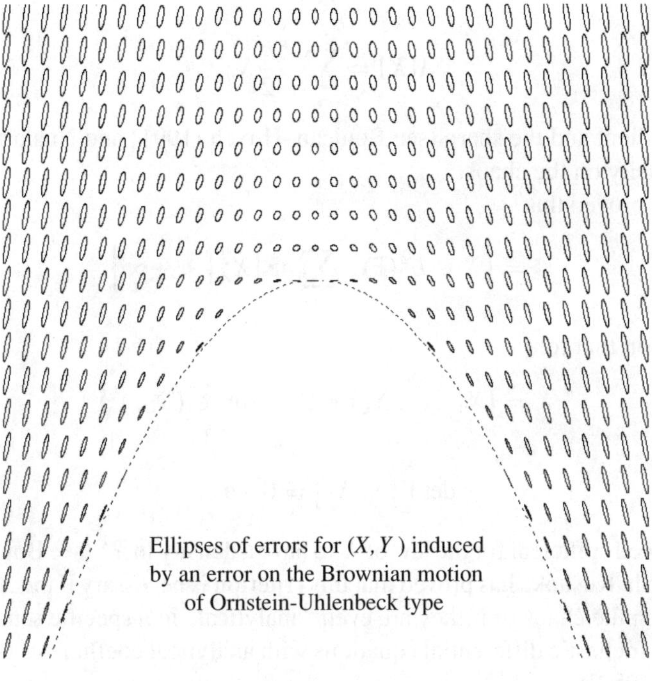

Ellipses of errors for (X, Y) induced
by an error on the Brownian motion
of Ornstein-Uhlenbeck type

Comment. The Ornstein–Uhlenbeck structure on the Wiener space is certainly the simplest error structure on this space and has been extensively studied. (see Bouleau–Hirsch [1991], Nualart [1995], Malliavin [1997], Üstünel and Zakai [2000], Ikeda–Watanabe [1981], etc.)

This structure is strongly related to the chaos decomposition of the Wiener space, discovered by Norbert Wiener and simplified by Kyiosi Itô using iterated Itô integrals.

Let h be a symmetric element of $L^2(\mathbb{R}_+^n, \lambda_n)$ (where λ_n is the Lebesgue measure in dimension n). If we consider

$$I_n(h) = n! \int_{0<t_1<\cdots<t_n} h(t_1, \ldots, t_n) dB_{t_1} \cdots dB_{t_n}$$

and denote C_n the subvector space of $L^2(\mathbb{P})$ spanned by such $I_n(h)$, the space $L^2(\mathbb{P})$ decomposes into a direct sum

$$L^2(\mathbb{P}) = \mathbb{R} \bigoplus_{n=1}^{\infty} C_n.$$

If $X = \sum_{n=0}^{\infty} X_n$ is the chaos expansion of $X \in L^2(\mathbb{P})$, $X \in \mathcal{D}A$ if and only if

$$\sum_n n^2 \mathbb{E}[X_n^2] < +\infty$$

and then

$$A[X] = \sum_n -\frac{n}{2} X_n.$$

Both the gradient and the sharp (see Bouleau–Hirsch [1991] and Nualart [1995]) are dealt with easily on the chaos.

It can be proved that

$$\mathbb{D} = \Big\{ X \in L^2(\mathbb{P}) \colon \sum_n n \mathbb{E}[X_n^2] < +\infty \Big\}.$$

Let us mention that for

$$X = \big(X_1, \ldots, X_k\big) \in \mathbb{D}^k \quad (\text{or } \in (\mathbb{D}_{\text{loc}})^k)$$

the criterion

$$\det \underline{\underline{\Gamma}}\big[X, X^t\big] \neq 0 \quad \text{a.e.}$$

has been proved sufficient for the law of X to have a density in $\mathbb{R}^k$ (see Bouleau–Hirsch [1986]) and Sh. Kusuoka has proved that this criterion is necessary if the X_k have finite expansions on the chaos or if they are even "analytical" in a specific sense, including solutions to stochastic differential equations with analytical coefficients (see Kusuoka [1981] and [1982]).

2.4 Structures with erroneous time. Let us now choose $(T, \mathcal{T}, \mu) = \big([0, 1], \mathcal{B}\big([0, 1]\big), dx\big)$ for the sake of simplicity and let

$$\begin{cases} \chi_n = \sqrt{2} \cos 2\pi n t & \text{if } n > 0 \\ \chi_0 = 1 \\ \chi_n = \sqrt{2} \sin 2\pi n t & \text{if } n < 0 \end{cases}$$

be the trigonometric basis of $L^2\big([0, 1]\big)$. We then follow the same construction as before:

$$B_t = \sum_n \int_0^t \chi_n(s) ds \cdot g_n$$

$$\int_0^1 f(s)\, dB_s = \sum_n \hat{f}_n g_n$$

if

$$f(t) = \sum_n \hat{f}_n \chi_n$$

and

$$(\Omega, \mathcal{A}, \mathbb{P}, \mathbb{D}, \Gamma) = \prod_n \big(\mathbb{R}, \mathcal{B}(\mathbb{R}), m, H^1(m), \gamma_n\big),$$

where m is the reduced normal law and

$$\gamma_n[u] = a_n u'^2$$

with a_n constant and dependent on n.

Example. $a_n = (2\pi n)^{2q}$, $q \in \mathbb{N}$. In this case

$$\Gamma\Big[\int_0^1 f(s)\,dB_s\Big] = \Gamma\Big[\sum_n \hat{f}_n g_n\Big] = \sum_n \hat{f}_n^2 (2\pi n)^{2q},$$

and from the theorem on products we know that

$$\int_0^1 f(s)\,dB_s \in \mathbb{D} \quad \text{if and only if} \quad \sum_n \hat{f}_n^2 (2\pi n)^{2q} < +\infty.$$

Proposition VI.13. *$\int_0^1 f(s)\,dB_s \in \mathbb{D}$ if and only if the q-th derivative $f^{(q)}$ of f in the sense of distribution belongs to $L^2([0,1])$; then*

$$\Gamma\Big[\int_0^1 f(s)\,dB_s\Big] = \int_0^1 f^{(q)2}(s)\,ds.$$

Proof. This result stems from the fact that $\big(f^{(q)} \in L^2([0,1])$ in the sense of $\mathcal{D}'\big)$ is equivalent with $\sum_n \hat{f}_n^2 (2\pi n)^{2q} < +\infty$, as easily seen using Fourier expansion. ◇

We can observe that the structure $(\Omega, \mathcal{A}, \mathbb{P}, \mathbb{D}, \Gamma)$ is *white* in the strong sense of error structures.

Proposition VI.14.

a) *Let $f \in L^2([0,1])$ with $f^{(q)} \in L^2([0,1])$ with such support that*

$$g = \tau_\alpha f = \big(t \to f(t-\alpha)\big)$$

also lies in $L^2([0,1])$. Then for $U = \int_0^1 f(s)\,dB_s$ and $V = \int_0^1 g(s)\,dB_s$, the image structures by both U and V are equal.

b) *Let $f, g \in L^2([0,1])$ and $f^{(q)}, g^{(q)} \in L^2([0,1])$ such that $fg = 0$, then for $U = \int_0^1 f(s)\,dB_s$ and $V = \int_0^1 g(s)\,dB_s$, the image structure by the pair (U, V) is the product of the image structures by U and by V.*

This result is also valid for the Ornstein–Uhlenbeck structure obtained for $q = 0$.

Proof. a) U and V display the same probability law as a centered Gaussian variable with variance $\|f\|^2_{L^2}$.

Let $\varphi \in \mathcal{C}^1 \cap \mathrm{Lip}$, the quadratic error operator of the image structure is

$$\Gamma_U[\varphi](x) = \mathbb{E}\big[\Gamma[\varphi \circ U] \mid U = x\big] = \varphi'^2(x) \int_0^1 g^{(q)2}(s)\, ds,$$

which is equal to $\Gamma_V[\varphi](x)$.

b) If $fg = 0$, U and V are independent Gaussian variables since the pair (U, V) is Gaussian and

$$\mathbb{E}[UV] = \int_0^1 f(s)g(s)\, ds = 0,$$

i.e.

$$(U, V)_*\mathbb{P} = (U_*\mathbb{P}) \times (V_*\mathbb{P}).$$

Let $\psi \in \mathcal{C}^1 \cap \mathrm{Lip}(\mathbb{R}^2)$

$$\Gamma\big[\psi(U, V)\big] = \psi_1'^2\Gamma[U] + 2\psi_1'\psi_2'\Gamma[U, V] + \psi_2'^2\Gamma[V].$$

We then apply the following fact, which has been left as an exercise.

Fact. If $f, g \in L^2([0, 1])$ are such that $f^{(q)}, g^{(q)} \in L^2([0, 1])$, then $fg = 0 \Rightarrow f^{(q)}g^{(q)} = 0$ a.e.

We then obtain

$$\Gamma[U, V] = \int_0^1 f^{(q)}g^{(q)}\, ds = 0,$$

hence

$$\Gamma_{(U,V)}[\psi](x, y) = \Gamma_U[\psi(\cdot, y)](x) + \Gamma_V[\psi(x, \cdot)](y). \qquad \diamond$$

Remark. In Proposition VI.14 the property of invariance by translation and that of independence in the sense of error structure are proved for random variables of the form $\int h(s)dB_s$, i.e. variables in the first chaos. These properties extend, mutatis mutandis, to general random variables by means of the functional calculus and the closedness property.

Exercise VI.15. Let us take $q = 1$.

a) Show that for $F \in \mathcal{C}^1 \cap \mathrm{Lip}(\mathbb{R}^k)$ and $\xi_i \in H^1([0, 1])$, the variable

$$X = F\left(\int_0^1 \xi_1\, dB, \dots, \int_0^1 \xi_k\, dB\right)$$

belongs to $\mathbb{D}$ and a gradient is defined by

$$D\left[\int_0^1 \xi_1\, dB\right] = \xi_1'.$$

b) Show that if $h \in H_0^1([0,1])$,

$$\langle DX, h\rangle = -\langle D_{OU}X, h'\rangle$$

where D_{OU} is the above-defined gradient of the Ornstein–Uhlenbeck structure.

c) For $F, G \in \mathbb{D} \cap L^\infty$ and $h \in H_0^1([0,1])$, prove

$$\mathbb{E}[G\langle DF, h\rangle] = -\mathbb{E}[F\langle DG, h\rangle] - \mathbb{E}\left[FG \int h'\, dB\right].$$

2.5 Structures of the generalized Mehler type. The error structures on the Wiener space constructed in the preceding Section 2.4 can be proved to belong to a more general family which will now be introduced.

Let $m = \mathcal{N}(0,1)$ as usual. Let us consider the probability space

$$(\Omega, \mathcal{A}, \mathbb{P}) = \big(\mathbb{R}, \mathcal{B}(\mathbb{R}), m\big)^{\mathbb{N}}$$

with g_n as coordinate mappings.

Let $X = F(g_0, \dots, g_n, \dots)$ be a bounded random variable. Consider the transform P_t:

$$P_t X = \hat{\mathbb{E}}\left[F\left(g_0\sqrt{e^{-a_0 t}} + \hat{g}_0\sqrt{1-e^{-a_0 t}}, \dots, g_n\sqrt{e^{-a_n t}} + \hat{g}_n\sqrt{1-e^{-a_n t}}, \dots\right)\right]$$

where the $\hat{g}_n$'s are copies of the g_n's, $\hat{\mathbb{E}}$ is the corresponding expectation and the a_n are positive numbers: $a_n \geq 0\ \forall n$.

The following properties are easily proved along the same lines as in dimension one (see Chapter II).

2.5.1. P_t is well-defined and preserves the probability measure $\mathbb{P}$.

2.5.2. P_t is continuous from $L^2(\mathbb{P})$ into itself with norm ≤ 1

$$\mathbb{E}\big(P_t X\big)^2 \leq \mathbb{E}P_t\big(X^2\big) = \mathbb{E}X^2.$$

2.5.3. P_t is a Markovian semigroup

$$\begin{aligned} P_{t+s}(X) &= P_t\big(P_s(X)\big) \\ P_t(X) &\geq 0 \quad \text{if } X \geq 0 \\ P_t(1) &= 1. \end{aligned}$$

2.5.4. P_t is symmetric with respect to $\mathbb{P}$.

Let $Y = G(g_0, \dots, g_n, \dots)$, we then obtain

$$\mathbb{E}\big[P_t X \cdot Y\big] = \mathbb{E}\big[F(\xi_0, \dots, \xi_n, \dots) G(y_0, \dots, y_n, \dots)\big]$$

where $\xi_0, \dots, \xi_n, \dots$ are i.i.d. reduced Gaussian variables and $y_0, \dots, y_n, \dots$ are also i.i.d. reduced Gaussian variables, such that $\operatorname{cov}(\xi_n, y_n) = \sqrt{e^{-a_n t}}$, i.e.

$$\mathbb{E}\big[P_t X \cdot Y\big] = \mathbb{E}\big[X \cdot P_t Y\big].$$

2.5.5. P_t is strongly continuous on $L^2(\mathbb{P})$. Indeed if X is bounded and cylindrical

$$\lim_{t \to 0} P_t X = X \quad \text{a.e.}$$

by virtue of dominated convergence, hence

$$\lim_{t \to 0} \mathbb{E}\big[(P_t X - X)^2\big] = 0$$

again by dominated convergence. From the density of bounded cylindrical random variables in $L^2(\mathbb{P})$, the result therefore follows.

2.5.6. Let us define

$$\mathbb{D} = \left\{ X \in L^2(\mathbb{P}) \colon \lim_{t \downarrow 0} \frac{1}{t} \mathbb{E}\big[(X - P_t X) X\big] < +\infty \right\}$$

and for $X \in \mathbb{D}$

$$\mathcal{E}[X] = \lim_{t \downarrow 0} \uparrow \frac{1}{t} \mathbb{E}\big[(X - P_t X) X\big].$$

By approximation on cylindrical functions, it can be shown that this construction provides the product error structure

$$(\Omega, \mathcal{A}, \mathbb{P}, \mathbb{D}, \Gamma) = \prod_{n=0}^{\infty} \big(\mathbb{R}, \mathcal{B}(\mathbb{R}), m, H^1(m), u \to a_n u'^2\big)$$

and

$$\mathbb{D} = \left\{ X = F(g_0, \dots, g_n, \dots) \colon \forall n \frac{\partial F}{\partial g_n} \in H^1(m) \ \sum_n a_n \left(\frac{\partial F}{\partial g_n}\right)^2 \in L^1(\mathbb{P}) \right\}$$

$$\Gamma[X] = \sum_n a_n \left(\frac{\partial F}{\partial g_n}\right)^2.$$

Let us now introduce the semigroup p_t on $L^2(\mathbb{R}_+)$ defined for

$$f = \sum_n \langle f, \chi_n \rangle \chi_n$$

by

$$p_t f = \sum_n \langle f, \chi_n \rangle e^{-a_n t} \chi_n.$$

(p_t) is a symmetric strongly continuous contraction semigroup on $L^2(\mathbb{R}_+)$ with eigenvectors χ_n. Let $(B, \mathcal{D}B)$ be its generator. Since

$$\|p_t f - f\|_{L^2}^2 = \sum_n \langle f, \chi_n \rangle^2 (1 - e^{-a_n t})^2$$

we can observe that if

$$\sum_n \langle f, \chi_n \rangle^2 a_n^2 < +\infty$$

then $f \in \mathcal{D}B$ and $Bf = -\sum_n \langle f, \chi_n \rangle a_n \chi_n$ which leads to

Proposition VI.16. *$\int_0^\infty f(s)\,dB_s \in \mathbb{D}$ if and only if*

$$\sum_n \langle f, \chi_n \rangle^2 a_n < +\infty,$$

i.e. using, in this case, the symbolic calculus notation

$$f \in \mathcal{D}(\sqrt{-B}), \quad \sqrt{-B} f = \sum_n \langle f, \chi_n \rangle \sqrt{a_n}\, \chi_n,$$

we then have

$$\Gamma\left[\int_0^\infty f(s)\,dB_s\right] = \langle \sqrt{-B} f, \sqrt{-B} f \rangle_{L^2(\mathbb{R}_+)}.$$

Proof. Since

$$\int_0^\infty f(s)\,dB_s = \sum_n \langle f, \chi_n \rangle g_n$$

and

$$\Gamma[g_n] = a_n, \quad \Gamma[g_m, g_n] = 0 \quad \text{if } m \neq n,$$

the result stems from both the theorem on products and the definition of $(B, \mathcal{D}B)$. ◇

Let us emphasize that the semigroup p_t on $L^2(\mathbb{R}_+)$ is not necessarily positive on positive functions. As a matter of fact, we obtained *any* symmetric, strongly continuous contraction semigroup on $L^2(\mathbb{R}_+)$, and we can start the construction with such a semigroup as input data.

Exercise VI.17. Show that for $f \in L^2(\mathbb{R}_+)$

$$P_t\left(\int_0^\infty f\,dB\right) = \int_0^\infty (p_{\frac{t}{2}} f)dB$$

$$P_t\left(\exp\left\{\int f\,dB - \frac{1}{2}\|f\|_{L^2}^2\right\}\right) = \exp\left\{\int p_{\frac{t}{2}} f\,dB - \frac{1}{2}\|p_{\frac{t}{2}} f\|_{L^2}^2\right\}$$

$$P_t\left(\left(\sin\int f\,dB\right)e^{\frac{1}{2}\|f\|_{L^2}^2}\right) = \left(\sin\int p_{\frac{t}{2}} f\,dB\right)e^{\frac{1}{2}\|p_{\frac{t}{2}} f\|_{L^2}^2}.$$

2.5.9. Considering the Wiener measure as carried by $\mathcal{C}_0(\mathbb{R}_+)$ and using the symbolic calculus for operators in $L^2(\mathbb{R}_+)$ the generalized Mehler formula can be demonstrated: $\forall F \in L^2(\Omega, \mathcal{A}, \mathbb{P})$:

$$P_t F = \hat{\mathbb{E}}\left[F\left(\int_0^\infty (p_{\frac{t}{2}} 1_{[0,\cdot]})(u)\,dB_u + \int_0^\infty (\sqrt{1-p_t}1_{[0,\cdot]})(v)\,d\hat{B}_v\right)\right].$$

This Mehler formula provides an intuitive interpretation of the error on the Brownian path modeled by this error structure. For example, in the Ornstein–Uhlenbeck case where $p_t u = e^{-t}u$, we can see that the path ω is perturbed in the following way:

$$\omega \longrightarrow e^{-\frac{\varepsilon}{2}}\omega + \sqrt{1-e^{\varepsilon}}\,\hat{\omega}$$

where $\hat{\omega}$ is an independent standard Brownian motion and ε a small parameter.

In the case of the weighted Ornstein–Uhlenbeck case (see Exercise VI.20 below)

$$\omega(s) = \int_0^s dB_u \longrightarrow \int_0^s e^{-\alpha(u)\varepsilon/2}\,dB_u + \int_0^s \sqrt{1-e^{-\alpha(u)\varepsilon}}\,d\hat{B}_u$$

(where α is a positive function in $L^1_{\text{loc}}(\mathbb{R}_+)$).

Example VI.18. Let $n(s,t)$ be a symmetric function in $L^2(\mathbb{R}_+^2, ds\,dt)$ and let us consider the operator from $L^2(\mathbb{R}_+)$ into $L^2(\mathbb{R}_+)$ defined by

$$L_n(f)(s) = \int_0^\infty n(s,t)f(t)\,dt.$$

a) Let $(u_i)_{i\in\mathbb{N}}$ be an orthonormal basis of $L^2(\mathbb{R}_+)$, then the quantity

$$\sum_i \|L_n(u_i)\|^2$$

does not depend on the basis (u_i) and is equal to

$$\|n\|^2_{L^2(\mathbb{R}_+^2, ds\,dt)}.$$

[Write $n(s,t) = \sum n_{ij} u_i(s) u_j(t)$.] The Hilbert–Schmidt norm of n is by definition

$$\|n\|_{HS} = \Big(\sum_i \|L_n(u_i)\|^2\Big)^{1/2} = \|n\|_{L^2(\mathbb{R}_+^2)}.$$

b) We obtain

$$\begin{aligned}\|L_n(f)\|^2 &= \int_0^\infty \Big(\int_0^\infty n(s,t) f(t)\,dt\Big)^2 ds \\ &\le \int_0^\infty \Big(\int_0^\infty (n(s,t))^2\,dt\Big)\Big(\int_0^\infty f^2(t)\,dt\Big)\,ds \\ &= \|L_n\|_{HS}^2 \|f\|_{L^2}^2.\end{aligned}$$

c) If we approximate $n(s,t)$ by

$$\sum_{i=1}^k y_i(s)\xi_i(t)$$

in $L^2(\mathbb{R}_+^2, ds\,dt)$, we can observe that the operator L_n is the limit for the operator norm of operators with finite dimensional range.

This statement implies that L_n is a *compact* operator (i.e. it maps the unit ball of $L^2(\mathbb{R}_+)$ into a relatively compact set).

By means of a famous theorem ascribed to Hilbert and Schmidt, the compact self-adjoint operator L_n possesses an orthonormal basis of eigenvectors, say v_i with eigenvalues c_i:

$$L_n v_i = c_i v_i$$

and

$$\|n\|_{L^2}^2 = \|L_n\|_{HS}^2 = \sum_i \|L_n(v_i)\|^2 = \sum_i |c_i|^2.$$

As semigroup p_t let us choose

$$p_t = e^{-(L_n)^2 t},$$

i.e.

$$p_t(v_i) = e^{-c_i^2 t} v_i \quad \forall i.$$

The above construction then yields the following result.

Proposition VI.19. *On the Wiener space* $(\Omega, \mathcal{A}, \mathbb{P}) = (\mathbb{R}, \mathcal{B}(\mathbb{R}), m)^{\mathbb{N}}$ *an error structure*

$$(\Omega, \mathcal{A}, \mathbb{P}, \mathbb{D}, \Gamma)$$

exists such that $\forall f \in L^2(\mathbb{R}_+)$, $\int_0^\infty f(s)\,dB \in \mathbb{D}$ *and*

$$\Gamma\Big[\int_0^\infty f(s)\,dB\Big] = \int_0^\infty \Big(\int_0^\infty n(s,t) f(t)\,dt\Big)^2 ds.$$

Exercise VI.20. Let α be a function on $\mathbb{R}_+$ such that $\alpha(x) \geq 0$ and $\alpha \in L^1_{\text{loc}}(\mathbb{R}_+, dx)$. Using the semigroup

$$p_t u = e^{-\alpha t} u,$$

show that an error structure on the Wiener space exists such that

$$\Gamma\left[\int_0^\infty f\, dB\right] = \int_0^\infty \alpha(x) f^2(x)\, dx$$

for f with the right-hand side being finite.

Exercise VI.21 (Application to Wiener filtering). In signal processing, Wiener filtering uses the Fourier transform, linear transformations and projections within the L^2-space of the spectral measure of the signal. Since the Fourier transform is, in practice, an erroneous mapping, it is only natural to consider stationary processes with erroneous spectral representation.

Consider two independent standard Brownian motions B^1_u, B^2_u, $u \geq 0$, and define

$$X_t = \int_0^\infty \frac{(\cos ut)\, dB^1_u + (\sin ut)\, dB^2_u}{\sqrt{1+u^2}}$$

for $t \in \mathbb{R}$. The process X_t is a Gaussian process, such that

$$\mathbb{E}\big[X_{t+s}\, X_s\big] = \int_0^\infty \frac{\cos ut}{1+u^2}\, du = \frac{\pi}{2} e^{-|t|}.$$

It is therefore stationary as a second-order stationary and centered Gaussian.

Let us set $\mathcal{F}_t = \sigma\big(X_s, s \leq t\big)$ and let $\mathcal{L}_t$ be the closure in $L^2(\mathbb{P})$ of the set of random variables $\sum_{i=1}^n a_i X_{t_i}$, $a_i \in \mathbb{R}$, $t_i \leq t$. Using the Gaussian character, $\mathbb{E}\big[X_{t+h} \mid \mathcal{F}_t\big]$ is easily shown to be the orthogonal projection of X_{t+h} on $\mathcal{L}_t$ in $L^2(\mathbb{P})$.

If $Z = \sum_{i=1}^n a_i X_{t_i}$, we have

$$\mathbb{E}\big[X_{t+h}\, Z\big] = \sum_i a_i \frac{\pi}{2} e^{-|t+h-t_i|}$$

and it follows without any difficulty that:

$$\mathbb{E}\big[X_{t+h} \mid \mathcal{F}_t\big] = e^{-h} X_t.$$

This result is not surprising once the real Ornstein–Uhlenbeck process, which is Markovian, has been recognized in $\big(X_t\big)$.

Let us introduce now an error on the spectral representation of $\big(X_t\big)$.

a) First of all let us assume B^1 and B^2 to be independently equipped with Ornstein–Uhlenbeck structures, so that for $f, g \in L^2(\mathbb{R}_+)$

$$\Gamma\left[\int_0^\infty f(u)\,dB_u^1\right] = \int_0^\infty f^2(u)\,du$$
$$\Gamma\left[\int_0^\infty g(u)\,dB_u^2\right] = \int_0^\infty g^2(u)\,du$$
$$\Gamma\left[\int_0^\infty f(u)\,dB_u^1, \int_0^\infty g(u)\,dB_u^2\right] = 0.$$

We can then note that $X_t \in \mathbb{D}\ \forall t \in \mathbb{R}$ and $\|X_t\|_{\mathbb{D}}^2 = 3\pi/4$. More generally,

$$\Big\|\sum_{i=1}^n a_i X_{t_i+h}\Big\|_{\mathbb{D}} = \Big\|\sum_{i=1}^n a_i X_{t_i}\Big\|_{\mathbb{D}}.$$

The process (X_t) is stationary for the norm $\|\cdot\|_{\mathbb{D}}$. The space $\mathcal{L}_t$ is closed in $\mathbb{D}$, and since $\|Z\|_{\mathbb{D}}^2 = \frac{3}{2}\|Z\|_{L^2}^2$ we observe that the forecasting problem has the same solution in the sense of $\|\cdot\|_{\mathbb{D}}$: $e^{-h}X_t$ is the projection of X_{t+h} on $\mathcal{L}_t$ in $\mathbb{D}$.

b) Next, let us take

$$\Gamma\left[\int_0^\infty f(u)\,dB_u^1\right] = \int_0^\infty \frac{f^2(u)}{(1+u^2)^2}\,du$$
$$\Gamma\left[\int_0^\infty g(u)\,dB_u^2\right] = \int_0^\infty \frac{g^2(u)}{(1+u^2)^2}\,du$$
$$\Gamma\left[\int_0^\infty f(u)\,dB_u^1, \int_0^\infty g(u)\,dB_u^2\right] = 0$$

for $f, g \in L^2(du)$.

The process (X_t) is still stationary for the norm $\|\cdot\|_{\mathbb{D}}$ but the norm $\|\cdot\|_{L^2}$ and $\|\cdot\|_{\mathbb{D}}$ are no longer equivalent.

If we introduce the stationary process Z_t of class $\mathcal{C}^2$ in $L^2(\mathbb{P})$ satisfying $Z_t - Z_t'' = X_t$ as given by

$$Z_t = \int_0^\infty \frac{\cos ut}{(1+u^2)^{3/2}}\,dB_u^1 + \int_0^\infty \frac{\sin ut}{(1+u^2)^{3/2}}\,dB_u^2,$$

it is verified that

$$\Gamma\big[X_{t+s}, X_s\big] = \mathbb{E}\big[Z_{t+s}, Z_s\big]$$

and

$$\Big\|\sum_{i=1}^n a_i X_{t_i}\Big\|_{\mathbb{D}}^2 = \Big\|\sum_{i=1}^n a_i X_{t_i}\Big\|_{L^2}^2 + \frac{1}{2}\Big\|\sum_{i=1}^n a_i Z_{t_i}\Big\|_{L^2}^2.$$

By these formulae, the process (Z_t) is a *gradient* for (X_t) with Hilbert space $L^2(\mathbb{P})$:

$$D[X_t] = Z_t.$$

c) Finally, let us consider an error structure on the spectral representation of X_t such that

$$\Gamma\left[\int_0^\infty f(u)\,dB_u^1\right] = \int_0^\infty f'^2(u)\,du$$
$$\Gamma\left[\int_0^\infty g(u)\,dB_u^2\right] = \int_0^\infty g'^2(u)\,du$$
$$\Gamma\left[\int_0^\infty f(u)\,dB_u^1, \int_0^\infty g(u)\,dB_u^2\right] = 0.$$

Such a structure is a generalized Mehler-type structure, with the semigroup p_t on $L^2(\mathbb{R}_+)$ being the semigroup of the Brownian motion with reflection at zero. In the above formulae f and g are in $H^1(\mathbb{R}_+)$.

We now have $X_t \in \mathbb{D}$ and

$$\Gamma[X_t] = \int_0^\infty \left(\frac{\partial}{\partial u}\frac{\cos ut}{\sqrt{1+u^2}}\right)^2 du + \int_0^\infty \left(\frac{\partial}{\partial u}\frac{\sin ut}{\sqrt{1+u^2}}\right)^2 du = t^2\frac{\pi}{2} + \frac{\pi}{16}.$$

We can note that (X_t) is no longer stationary for the $\mathbb{D}$-norm: a differential in the frequency domain is too strong of a perturbation to maintain the stationarity. $\Gamma[X_t]$ is minimal for $t = 0$ and

$$\lim_{|t|\to\infty} \frac{\sqrt{\Gamma[X_t]}}{t} = \sqrt{\frac{\pi}{2}},$$

asymptotically, the standard deviation of the error increases linearly with time.

Remark. In the case of the Wiener space or Gaussian stationary processes, the Gaussian techniques allow generalizing the processes and random variables to distributions in suitable sense: white noise theory (see S. Watanabe [1984], J. Potthoff [1987], D. Feyel and A. de La Pradelle [1989]).

Exercise (Image of a generalized Mehler-type error structure by the Itô application of an S.D.E.). Let us consider a symmetric strongly continuous contraction semigroup (p_t) on $L^2(\mathbb{R}_+)$ and set

$$Z_t^\varepsilon = \int_0^\infty \left(p_{\frac{\varepsilon}{2}} 1_{[0,t]}\right)(u)\,dB_u + \int_0^\infty \left(\sqrt{I - p_\varepsilon}\, 1_{[0,t]}\right)(v)\,d\hat{B}_v,$$

for fixed ε, $\left(Z_t^\varepsilon\right)_{t\geq 0}$ is a standard Brownian motion on $\mathbb{R}_+$ defined on the probability space $(\Omega, \mathcal{A}, \mathbb{P}) \times (\hat{\Omega}, \hat{\mathcal{A}}, \hat{\mathbb{P}})$.

Let σ and b be Lipschitzian applications from $\mathbb{R}$ to $\mathbb{R}$ and $\left(X_t^\varepsilon\right)_{t\geq 0}$ be the solution of

$$X_t^\varepsilon = x + \int_0^t \sigma\left(X_s^\varepsilon\right) dZ_s^\varepsilon + \int_0^t b\left(X_s^\varepsilon\right) ds.$$

We then define a linear operator Q_ε on the space of bounded measurable functions from $\mathcal{C}\left([0,\infty[\right)$ into $\mathbb{R}$ by

$$\left(Q_\varepsilon[G]\right)(\xi) = \mathbb{E}\left\{\hat{\mathbb{E}}\left[G(X^\varepsilon)\right] | X_\cdot^0 = \xi\right\}$$

for any G bounded and measurable from $\mathcal{C}([0,\infty[)$ into $\mathbb{R}$.

We denote ψ the Itô's map which to $\mathbb{P}$-a.e. $\omega \in \Omega$ associates the solution $\left(X_t^0\right)_{t\geq 0}$ of

$$X_t^0 = x + \int_0^t \sigma\left(X_s^0\right) dB_s + \int_0^t b\left(X_s^0\right) ds.$$

Let $(\Omega, \mathcal{A}, \mathbb{P}, \mathbb{D}, \Gamma)$ be the error structure of the generalized Mehler type, $(P_\varepsilon)_{\varepsilon\geq 0}$ be its semi-group defined in Section 2.5 and ν be the law of $X_\cdot^0$, i.e. the probability measure on $\mathcal{C}([0,\infty[)$ image of the Wiener measure by ψ.

Although we cannot assert that the family $(Q_\varepsilon)_{\varepsilon\geq 0}$ is a semi-group (ψ is not necessarily one-to-one outside a negligible set), the following properties can be verified:

a) $\forall F, G \in L^2(\nu)$

$$\langle Q_\varepsilon[F], G\rangle_\nu = \langle P_\varepsilon[F\circ\psi], G\circ\psi\rangle_{\mathbb{P}}$$

and the operator Q_ε is symmetric on $L^2(\nu)$.

b) The family (Q_ε) is strongly continuous in $L^2(\nu)$: $\forall F \in L^2(\nu)$,

$$\begin{aligned} &\lim\nolimits_{\varepsilon\to 0} Q_\varepsilon[F] = F \\ &\lim\nolimits_{\varepsilon\to\varepsilon_0} Q_\varepsilon[F] = Q_{\varepsilon_0}[F]. \end{aligned}$$

c)

$$0 \leq \frac{1}{\varepsilon}\langle F - Q_\varepsilon[F], F\rangle \quad \forall F \in L^2(\nu).$$

d)

$$\left\{F \in L^2(\nu) : \lim_{\varepsilon\to 0} \frac{1}{\varepsilon}\langle F - Q_\varepsilon[F], F\rangle_\nu < +\infty\right\} = \left\{F \in L^2(\nu) : F\circ\psi \in \mathbb{D}\right\}.$$

Let us denote this set by $\tilde{\mathbb{D}}$.

e) If $\tilde{\mathbb{D}} = \{F \in L^2(\nu) \;:\; F\circ\psi \in \mathbb{D}\}$ is dense in $L^2(\nu)$, we then set for $F \in \tilde{\mathbb{D}}$

$$\tilde{\Gamma}[F](\xi) = \mathbb{E}[\Gamma[F\circ\psi] | \psi = \xi].$$

Then

$$\left(\mathcal{C}([0,\infty[), \mathcal{B}(\mathcal{C}([0,\infty[)), \nu, \tilde{\mathbb{D}}, \tilde{\Gamma}\right)$$

is an error structure whose Dirichlet form

$$\tilde{\mathcal{E}}[F] = \frac{1}{2}\int \tilde{\Gamma}[F]\,d\nu$$

is given by

$$\tilde{\mathcal{E}}[F] = \lim_{\varepsilon\to 0} \frac{1}{\varepsilon}\langle F - Q_\varepsilon F, F\rangle_\nu .$$

3 Error structures on the Poisson space

Several error structures can easily be constructed either on the Poisson process on $\mathbb{R}_+$ or on the general Poisson space. As in the case of Brownian motion, these structures allow studying more sophisticated objects, such as marked processes and processes with independent increments, which can be defined in terms of a general Poisson point process.

Among the works on the variational calculus on the Poisson space let us first cite Bichteler–Gravereau–Jacod [1987] and Wu [1987]. The construction produced by these authors yields the same objects as our approach in Section 3.2. Carlen and Pardoux, in 1990, introduced a different structure on the Poisson process on $\mathbb{R}_+$ and displayed some interesting properties. This domain represents still an active field of research (Nualart and Vives [1990], Privault [1993], Decreusefond [1998], etc.).

Our initial approach will consist of following to the greatest extent possible the classical construction of a Poisson point process, which we will first recall:

3.1 Construction of a Poisson point process with intensity measure μ. Let us begin with the case where μ is a finite measure.

3.1.1. Let $(G, \mathcal{G}, \mu)$ be a measurable space equipped with a finite positive measure μ. We set $\theta = \mu(G)$ and $\mu_0 = \frac{1}{\theta}\cdot\mu$. Considering the product probability space

$$(\Omega, \mathcal{A}, \mathbb{P}) = \big(G, \mathcal{G}, \mu_0\big)^{\mathbb{N}^*} \times \big(\mathbb{N}, \mathcal{P}(\mathbb{N}), P_\theta\big),$$

where $\mathcal{P}(\mathbb{N})$ denotes the σ-field of all subsets of integers $\mathbb{N}$ and P_θ denotes the Poisson law on $\mathbb{N}$ with parameter θ defined by

$$P_\theta(\{n\}) = e^{-\theta}\frac{\theta^n}{n!}, \qquad n \in \mathbb{N},$$

and if we denote the coordinate mappings of this product space by $\big(X_n\big)_{n>0}$ and Y, we obtain for the X_n's a sequence of random variables with values in $(G, \mathcal{G})$ which are i.i.d. with law μ_0 and for Y an integer-valued random variable with law P_θ independent of the X_n's.

The following formula

$$N(\omega) = \sum_{n=1}^{Y(\omega)} \delta_{X_n(\omega)},$$

where δ is the Dirac measure (using the convention $\sum_1^0 = 0$) defines a random variable with values in the space of "point measures", i.e. measures which are sum of Dirac measures. Such a random variable is usually called a "point process."

Proposition VI.22. *The point process N features the following properties:*

a) *If $A_1, \ldots, A_n$ are in $\mathcal{G}$ and pairwise disjoint then the random variables $N(A_1), \ldots, N(A_n)$ are independent.*

b) *For $A \in \mathcal{G}$, $N(A)$ follows a Poisson law with parameter $\mu(A)$.*

Proof. This result is classical (see Neveu [1977] or Bouleau [2000]). ◇

Since the expectation of a Poisson variable is equal to the parameter, we have $\forall A \in \mathcal{G}$

$$\mu(A) = \mathbb{E}\big[N(A)\big]$$

such that μ can be called the *intensity* of point process N.

3.1.2. Let us now assume that the space $(G, \mathcal{G}, \mu)$ is only σ-finite. A sequence $G_k \in \mathcal{G}$ then exists such that

- the G_k are pairwise disjoint,
- $\bigcup_k G_k = G$,
- $\mu\big(G_k\big) < +\infty$.

Let us denote $\big(\Omega_k, \mathcal{A}_k, \mathbb{P}_k\big)$ and N_k the probability spaces and point processes obtained by the preceding procedure on G_k; moreover let us set

$$(\Omega, \mathcal{A}, \mathbb{P}) = \prod_k \big(\Omega_k, \mathcal{A}_k, \mathbb{P}_k\big)$$

$$N = \sum_k N_k.$$

We then obtain the same properties for N as in Proposition VI.23, once the parameters of the Poisson laws used are finite.

Such a random point measure is called a Poisson point process with intensity μ.

3.1.3. Let us indicate the Laplace characteristic functional of N.

For $f \geq 0$ and $\mathcal{G}$-measurable

$$\mathbb{E}e^{-N(f)} = \exp\left\{-\int \left(1 - e^{-f}\right) d\mu\right\}.$$

3.2 The white error structure on the general Poisson space. The first error structure that we will consider on the Poisson space displays the property that each point thrown in space G is erroneous and modeled by the same error structure on $(G, \mathcal{G})$, moreover if we examine the points in A_1 and their errors along with the points in A_2 and their errors, there is independence if $A_1 \cap A_2 = \emptyset$. This property justifies the expression "white error structure".

3.2.1. Let us begin with the case where μ is finite. Suppose an error structure is given on $\left(G, \mathcal{G}, \mu_0\right)$ e.g. $\left(G, \mathcal{G}, \mu_0, \boldsymbol{d}, \gamma\right)$; using the theorem on products once more, if we set

$$(\Omega, \mathcal{A}, \mathbb{P}, \mathbb{D}, \Gamma) = \left(G, \mathcal{G}, \mu_0, \boldsymbol{d}, \gamma\right)^{\mathbb{N}^*} \times \left(\mathbb{N}, \mathcal{P}(\mathbb{N}), P_\theta, L^2(P_\theta), 0\right),$$

we obtain an error structure that is Markovian if $(G, \mathcal{G}, \mu_0, \boldsymbol{d}, \gamma)$ is Markovian.

Then any quantity depending on

$$N = \sum_{n=1}^{Y} \delta_{X_n}$$

and sufficiently regular will be equipped with a quadratic error:

Proposition VI.23. *Let* $U = F\left(Y, X_1, X_2, \ldots, X_n, \ldots\right)$ *be a random variable in* $L^2(\Omega, \mathcal{A}, \mathbb{P})$, *then*

a) $U \in \mathbb{D}$ *if and only if* $\forall m \in \mathbb{N}, \forall k \in \mathbb{N}^*$, *for* $\mu_0^{\otimes \mathbb{N}^*}$*-a.e.* $x_1, \ldots, x_{k-1}, x_{k+1}, \ldots$

$$F\left(m, x_1, \ldots, x_{k-1}, \cdot, x_{k+1}, \ldots\right) \in \boldsymbol{d}$$

and $\mathbb{E}\left[\sum_{k=1}^{\infty} \gamma_k[F]\right] < +\infty$ *(where, as usual,* γ_k *is* γ *acting upon the k-th variable);*

b) *for* $U \in \mathbb{D}$

$$\Gamma[U] = \sum_{k=1}^{\infty} \gamma_k\left[F\left(Y, X_1, \ldots, X_{k-1}, \cdot, X_{k+1}, \ldots\right)\right]\left(X_k\right).$$

Proof. This is simply the theorem on products. ◇

This setting leads to the following proposition:

Proposition VI.24. *Let $f, g \in \boldsymbol{d}$, then $N(f)$ and $N(g)$ are in $\mathbb{D}$ and*

$$\Gamma\big[N(f)\big] = N\big(\gamma[f]\big)$$
$$\Gamma\big[N(f), N(g)\big] = N\big(\gamma[f, g]\big).$$

Proof. By $\mathbb{E}|N(f) - N(g)| \le \mathbb{E}[N|f-g|] = \mu|f-g|$, the random variable $N(f)$ depends solely upon the μ-equivalence class of f.

From the Laplace characteristic functional, we obtain

$$\mathbb{E}\big[N(f)^2\big] = \int f^2\,d\mu + \Big(\int f\,d\mu\Big)^2,$$

thus proving that $N(f) \in L^2(\mathbb{P})$ if $f \in L^2(\mu)$. Then for $f \in \boldsymbol{d}$,

$$\begin{aligned}\Gamma\big[N(f)\big] &= \sum_{k=1}^{\infty} \gamma_k \left[\sum_{n=1}^{Y} f(X_n)\right] = \sum_{k=1}^{\infty} 1_{\{k \le Y\}} \gamma[f](X_k)\\ &= \sum_{k=1}^{Y} \gamma[f](X_k) = N\big(\gamma[f]\big).\end{aligned}$$

The required result follows. ◇

By functional calculus, this proposition allows computing Γ on random variables of the form $F\big(N(f_1), \dots, N(f_k)\big)$ for $F \in \mathcal{C}^1 \cap \mathrm{Lip}$ and $f_i \in \boldsymbol{d}$.

Let $(a, \mathcal{D}a)$ be the generator of the structure $\big(G, \mathcal{G}, \mu_0, \boldsymbol{d}, \gamma\big)$, we also have:

Proposition VI.25. *If $f \in \mathcal{D}a$, then $N(f) \in \mathcal{D}A$ and*

$$A\big[N(f)\big] = N\big(a[f]\big).$$

Proof. The proof is straightforward from the definition of N. ◇

For example if $f \ge 0$, $f \in \mathcal{D}a$, then

$$A\big[e^{-\lambda N(f)}\big] = e^{-\lambda N(f)} N\left(\frac{\lambda^2}{2}\gamma[f] - \lambda a[f]\right).$$

3.2.2. Chaos. Let us provide some brief comments on the chaos decomposition of the Poisson space. Let us set $\tilde{N} = N - \mu$. If $A_1, \dots, A_k$ are pairwise disjoint sets in $\mathcal{G}$, we define

$$I_k\big(1_{A_1} \otimes \cdots \otimes 1_{A_k}\big) = \tilde{N}(A_1) \cdots \tilde{N}(A_k),$$

the operator I_k extends uniquely to a linear operator on $L^2\big(G^k, \mathcal{G}^{\otimes k}, \mu^k\big)$ such that

- $I_k(f) = I_k(\tilde{f})$, where $\tilde{f}$ is the symmetrized function of f,
- $\mathbb{E} I_k(f) = 0\ \forall k \geq 1$, $I_0(f) = \int f\, d\mu$,
- $\mathbb{E}\big[I_p(f) I_q(g)\big] = 0$ if $p \neq q$,
- $\mathbb{E}\big[(I_p(f))^2\big] = p! \langle \tilde{f}, \tilde{g} \rangle_{L^2(\mu^p)}$.

If C_n is the subvector space of $L^2(\Omega, \mathcal{A}, \mathbb{P})$ of $I_n(f)$, we then have the direct sum

$$L^2(\Omega, \mathcal{A}, \mathbb{P}) = \bigoplus_{n=0}^{\infty} C_n.$$

The link of the white error structure on the Poisson space with the chaos decomposition is slightly analogous to the relation of generalized Mehler-type error structures with the chaos decomposition on the Wiener space. It can be shown that if (P_t) is the semigroup on $L^2(\mathbb{P})$ associated with error structure $(\Omega, \mathcal{A}, \mathbb{P}, \mathbb{D}, \Gamma)$, then $\forall f \in L^2(G^n, \mathcal{G}^{\otimes n}, \mu^n)$

$$P_t\big(I_n(f)\big) = I_n\big(p_t^{\otimes n} f\big),$$

where (p_t) is the semigroup on $L^2(\mu_0)$ associated with the error structure $\big(G, \mathcal{G}, \mu_0, \boldsymbol{d}, \gamma\big)$.

It must nevertheless be emphasized that p_t here is necessarily positive on positive functions whereas this condition was not compulsory in the case of the Wiener space.

Exercise VI.26. Let d be a gradient for $\big(G, \mathcal{G}, \mu_0, \boldsymbol{d}, \gamma\big)$ with values in the Hilbert space H. Let us define $\mathcal{H}$ by the direct sum

$$\mathcal{H} = \bigoplus_{n=1}^{\infty} H_n,$$

where H_n are copies of H.

Show that for $U = F\big(Y, X_1, \ldots, X_n, \ldots\big) \in \mathbb{D}$,

$$D[U] = \sum_{k=1}^{\infty} d_k\big[F\big(Y, X_1, \ldots, X_{k-1}, \cdot, X_{k+1}, \ldots\big)\big](X_k)$$

defines a gradient for $(\Omega, \mathcal{A}, \mathbb{P}, \mathbb{D}, \Gamma)$.

3.2.3 σ-finite case. When μ is σ-finite, the construction may be performed in one of several manners which do not all yield the same domains.

If we try to strictly follow the probabilistic construction (see Section 3.1.2) it can be assumed that we have error structures on each G_k

$$S_k = \Big(G_k, \mathcal{G}\big|_{G_k}, \frac{1}{\mu(G_k)} \mu\Big|_{G_k}, \boldsymbol{d}_k, \gamma_k\Big)$$

hence, as before, we have error structures on $\big(\Omega_k, \mathcal{A}_k, \mathbb{P}_k\big)$, e.g. $\big(\Omega_k, \mathcal{A}_k, \mathbb{P}_k, \mathbb{D}_k\Gamma_k\big)$, and Poisson point processes N_k.

We have noted that on

$$(\Omega, \mathcal{A}, \mathbb{P}) = \prod_k (\Omega_k, \mathcal{A}_k, \mathbb{P}_k)$$
$$N = \sum_k N_k$$

is a Poisson point process with intensity μ. Thus, it is natural to take

$$(\Omega, \mathcal{A}, \mathbb{P}, \mathbb{D}, \Gamma) = \prod_k \big(\Omega_k, \mathcal{A}_k, \mathbb{P}_k, \mathbb{D}_k, \Gamma_k\big).$$

Let us define

$$\underline{\boldsymbol{d}} = \big\{f \in L^2(\mu) \colon \forall k \ f|_{G_k} \in \boldsymbol{d}_k\big\}$$

and for all $f \in \underline{\boldsymbol{d}}$, let us set

$$\gamma[f] = \sum_k \gamma_k\big[f|_{G_k}\big].$$

We then have the following result.

Proposition VI.27. *Let $f \in \underline{\boldsymbol{d}}$ be such that $f \in L^1 \cap L^2(\mu)$ and $\gamma[f] \in L^1(\mu)$. Then $N(f) \in \mathbb{D}$ and*

$$\Gamma\big[N(f)\big] = N\big(\gamma[f]\big).$$

Proof. $N(f)$ is defined because

$$\mathbb{E}N(|f|) = \int |f|\, d\mu < +\infty$$
$$N(f) = \sum_k N_k\big(f|_{G_k}\big)$$

and from the theorem on products, $N(f) \in \mathbb{D}$ and

$$\Gamma\big[N(f)\big] = \sum_k \Gamma_k\big[N_k\big(f|_{G_k}\big)\big] = \sum_k N_k\big(\gamma_k\big[f|_{G_k}\big]\big) = N\big(\gamma[f]\big). \qquad \diamond$$

To clearly see what happens with the domains, let us proceed with the particular case where

$$(G, \mathcal{G}) = \big(\mathbb{R}_+, \mathcal{B}(\mathbb{R}_+)\big),$$

μ is the Lebesgue measure on $\mathbb{R}_+$, G_k are the intervals $[k, k+1[$, and the error structures S_k are

$$\big([k, k+1[, \mathcal{B}([k, k+1[), dx, H^1([k, k+1[), u \to u'^2\big).$$

We then have in $\underline{\boldsymbol{d}}$ not only continuous functions with derivatives in $L^2_{\text{loc}}(dx)$, but also discontinuous functions with jumps at the integers.

Practically, this is not troublesome. We thus have

Lemma. *The random σ-finite measure*

$$\tilde{N} = N - \mu$$

extends uniquely to $L^2(\mathbb{R}_+)$ and for $f \in H^1(\mathbb{R}_+, dx)$

$$\Gamma[\tilde{N}(f)] = N(f'^2).$$

Proof. The first property is derived from

$$\mathbb{E}\left(N(f) - \int f\,dx\right)^2 = \int f^2\,dx \quad \text{for } f \in \mathcal{C}_K(\mathbb{R}_+).$$

The second stems from the above construction because $H^1(\mathbb{R}_+, dx) \subset \underline{\boldsymbol{d}}$. ◇

3.2.4 Application to the Poisson process on $\mathbb{R}_+$. Let us recall herein our notation.

On $[k, k+1[$, we have an error structure

$$S_k = \big([k, k+1[, B([k, k+1[), dx, H^1([k, k+1[), u \xrightarrow{\gamma_k} u'^2\big).$$

With these error structures, we built Poisson point processes on $[k, k+1[$ and then placed error structures on them:

$$\big(\Omega_k, \mathcal{A}_k, \mathbb{P}_k, \mathbb{D}_k, \Gamma_k\big) = \big(\mathbb{N}, \mathcal{P}(\mathbb{N}), P_1, L^2(P_1), 0\big) \times \big(S_k\big)^{\mathbb{N}^*}.$$

If $Y^k, X_1^k, X_2^k, \dots, X_n^k, \dots$ denote the coordinate maps, the point process is defined by

$$N^k = \sum_{n=1}^{Y^k} \delta_{X_n^k}.$$

We have proved that for $f \in H^1([k, k+1[)$

$$\Gamma_k[N^k(f)] = N^k(f'^2)$$

and for $f \in \mathcal{C}^2([k, k+1])$ with $f'(k) = f'(k+1) = 0$,

$$A_k[N^k(f)] = \frac{1}{2} N^k(f'').$$

(cf. Example III.3 and Propositions VI.24 and VI.25).

We now take the product

$$(\Omega, \mathcal{A}, \mathbb{P}, \mathbb{D}, \Gamma) = \prod_{k=0}^{\infty} (\Omega_k, \mathcal{A}_k, \mathbb{P}_k, \mathbb{D}_k, \Gamma_k)$$

and set

$$N = \sum_{k=0}^{\infty} N^k.$$

Let us denote ξ_k the coordinate mappings of this last product, we then have from the theorem on products

Lemma VI.28.

- $\forall k \in \mathbb{N}, \forall n \in \mathbb{N}^*, X_n^k \circ \xi_k \in \mathbb{D}$,
- $\Gamma\big[X_n^k \circ \xi_k\big] = 1$,
- $\Gamma\big[X_m^k \circ \xi_k, X_n^\ell \circ \xi_\ell\big] = 0$ *if* $k \neq \ell$ *or* $m \neq n$.

If we set $N_t = N\big([0, t]\big)$, N_t is a usual Poisson process with unit intensity on $\mathbb{R}_+$. Let $T_1, T_2, \ldots, T_i, \ldots$ be its jump times.

We can prove

Proposition VI.29. *T_i belongs to $\mathbb{D}$.*

$$\Gamma\big[T_i\big] = 1, \quad \Gamma\big[T_i, T_j\big] = 0 \;\; \textit{if } i \neq j.$$

Proof. We will exclude ξ_k in the notation for the sake of simplicity.

If $Y^0 \geq 1$, there is at least one point in $[0, 1[$, T_1 is defined by

$$T_1 = \inf\big\{X_n^0, n = 1, 2, \ldots, Y^0\big\}.$$

If $Y^0 = 0$ and $Y^1 \geq 1$, T_1 is defined by

$$T_1 = \inf\big\{X_n^1, n = 1, 2, \ldots, Y^1\big\}$$

etc. In any case we have

$$T_1 = \sum_{k=0}^{\infty} \Big(\bigwedge_{n=1}^{Y^k} X_n^k\Big) 1_{\{Y^0 = Y^1 = \cdots = Y^{k-1} = 0 < Y^k\}}.$$

The sets

$$\big\{Y^0 = Y^1 = \cdots = Y^{k-1} = 0 < Y^k\big\}$$

are disjoint with union Ω and depend solely on the Y's which are not erroneous. The random variables

$$\bigwedge_{n=1}^{Y^k} X_n^k$$

are in $\mathbb{D}$ since Lipschitz functions operate. Finally

$$\begin{aligned}\Gamma\big[T_1\big] &= \sum_k \Gamma\Big[\bigwedge_{n=1}^{Y^k} X_n^k\Big] 1_{\{Y^0=\cdots=Y^{k-1}=0<Y^k\}}\\ &= 1\end{aligned}$$

because $\Gamma\big[X_n^k\big] = 1$ and because of Proposition III.15.

The argument for $\Gamma\big[T_i\big]$ and $\Gamma\big[T_i, T_j\big]$ is similar, only the notation is more sophisticated. Let us write T_2, for example:

$$T_2 = \sum_{k=0}^{\infty}\Big(\bigwedge_{n=1}^{Y^k} X_n^k\Big) 1_{\{Y^0+\cdots+Y^{k-1}=1,\, Y^k=1\}} + \Big(\bigvee_n \bigwedge_{\substack{m=1,\cdots,Y^k\\ m\neq n}} X_n^k\Big) 1_{\{Y^0+\cdots+Y^{k-1}=0,\, Y^k=2\}}.$$

We note in the same manner that $\Gamma[T_2] = 0$ and for $\Gamma[T_1, T_2]$, we observe that only $\Gamma[X_n^k, X_{n'}^{k'}]$ will appear in the calculation with $n \neq n'$ or $k \neq k'$.

This leads to the result. ◇

Corollary VI.30.

a) *If F is $\mathcal{C}^1 \cap \mathrm{Lip}$ then*

$$\Gamma\big[F\big(T_1, \ldots, T_p\big)\big] = \sum_{i=1}^{p} F_i'^2\big(T_1, \ldots, T_p\big).$$

b) *For $f \in H^1(\mathbb{R}_+)$, $\int_0^\infty f(s)\, d(N_s - s) \in \mathbb{D}$ and*

$$\Gamma\Big[\int_0^\infty f(s)\, d(N_s - s)\Big] = \int_0^\infty f'^2(s)\, dN_s.$$

c) *For $f \in H^1(\mathbb{R}_+)$ with $f'(0) = 0$ and $f'' \in L^1(\mathbb{R}_+) \cap L^2(\mathbb{R}_+)$ we have*

$$\int_0^\infty f(s)\, d(N_s - s) \in \mathcal{D}A$$

and

$$A\Big[\int_0^\infty f(s)\, d(N_s - s)\Big] = \frac{1}{2}\int_0^\infty f''(s)\, dN_s.$$

Proof. Let us present the argument for point c), as an example.

As usual we must find a v, such that

$$\frac{1}{2}\mathbb{E}\Gamma\left[\int_0^\infty f(s)\,d(N_s - s), X\right] = \mathbb{E}[vX]$$

for $X = F(T_1, \dots, T_n)$, $F \in \mathcal{C}^1 \cap \mathrm{Lip}$. The left-hand member is also

$$\frac{1}{2}\sum_{i=1}^n \mathbb{E}\big[f'(T_i)F_i'(T_1, \dots, T_n)\big]$$

Considering first the following decomposition:

$$\begin{aligned}\mathbb{E}\sum_{i=1}^n f'(T_i)F_i'(T_1, \dots, T_n) = \mathbb{E}\big[&f'(T_1)(F_1' + \cdots + F_n') \\ &+ (f'(T_2) - f'(T_1))(F_2' + \cdots + F_n') \\ &+ \cdots \\ &+ (f'(T_i) - f'(T_{i-1}))(F_i' + \cdots + F_n') \\ &+ \cdots \\ &+ (f'(T_n) - f'(T_{n-1}))F_n'\big].\end{aligned}$$

Then calculating the term $\mathbb{E}[(f'(T_i) - f'(T_{i-1}))(F_i' + \cdots + F_n')]$ by means of an integration by parts on the exponential variable E_i, where $T_n = E_1 + \cdots + E_n$, yields

$$\begin{aligned}\mathbb{E}\textstyle\sum_{i=1}^n f'(T_i)F_i'(T_1, \dots, T_n) = -f'(0)\mathbb{E}[F(0, T_1, \dots, T_{n-1})] + \mathbb{E}[f'(T_n)X] \\ -\mathbb{E}[(f''(T_1) + \cdots + f''(T_n))X].\end{aligned}$$

Using now the easy fact that for $h \in L^1(\mathbb{R}_+)$

$$\mathbb{E}\left[\int_0^\infty h(s)\,dN_s \mid T_1, \dots, T_n\right] = h(T_1) + \cdots + h(T_n) + \int_{T_n}^\infty h(s)\,ds,$$

we obtain, since $f'(0) = 0$,

$$\mathbb{E}\sum_{i=1}^n f'(T_i)F_i'(T_1, \dots, T_n) = -\mathbb{E}[N(f'')X].$$

We may thus take $v = \frac{1}{2}N(f'')$ and the statement follows using the closedness of operator A. ◇

3.2.5 Application to internalization. The construction discussed above is indispensable for studying random variables that depend on an infinite number of T_n.

Nevertheless, it also gives results in finite dimension, which could be elementarily proved using the fact that random variables $T_{n+1} - T_n$ are i.i.d. with exponential law. We have, for instance, the following results.

Lemma. *Let $g \in \mathcal{C}^1(\mathbb{R}_+)$ with polynomial growth and vanishing at zero. Let $F \in \mathcal{C}^1 \cap \operatorname{Lip}(\mathbb{R}^n)$. Then*

$$\mathbb{E}\Big[\sum_{i=1}^{n} g(T_i)F_i'(T_1, \dots, T_n)\Big] = \mathbb{E}\Big[\Big(g(T_n) - \sum_{i=1}^{n} g'(T_i)\Big)F(T_1, \dots, T_n)\Big].$$

Proof. Let us first consider an f as in Corollary VI.30. The proof of this corollary yields

$$\mathbb{E}\Big[\sum_{i=1}^{n} f'(T_i)F_i'(T_1, \dots, T_n)\Big] = \mathbb{E}\Big[\Big(f'(T_n) - \sum_{i=1}^{n} f''(T_i)\Big)F(T_1, \dots, T_n)\Big].$$

This relation now extends to the hypotheses of the statement by virtue of dominated convergence. ◇

With the same hypotheses on F, the lemma directly yields the following formula

$$\frac{d}{d\alpha}\mathbb{E}\big[F\big(\alpha T_1, \dots, \alpha T_n\big)\big] = \frac{1}{\alpha}\mathbb{E}\big[(T_n - n)F\big(\alpha T_1, \dots, \alpha T_n\big)\big]. \tag{5}$$

Exercise. Provide a formula without derivation for

$$\frac{d}{d\alpha}\mathbb{E}\big[F\big(\alpha h(T_1), \dots, \alpha h(T_n)\big)\big].$$

Exercise. Consider the random variable with values in $\mathbb{R}^2$ $X = \big(N(f_1), N(f_2)\big)$ for $f_1, f_2 \in L^1 \cap L^2(\mathbb{R}_+)$; show that if

$$\det \Gamma\big[N(f_i), N(f_j)\big] = 0 \quad \mathbb{P}\text{-a.s.},$$

then the law of X is carried by a straight line.

3.3 The Carlen–Pardoux error structure. For the classical Poisson process on $\mathbb{R}_+$, E. Carlen and E. Pardoux have proposed and studied an error structure which possesses a gradient and a δ with attractive properties.

As previously mentioned, if T_n are the jump times of the Poisson process, random variables $E_n = T_n - T_{n-1}$, $n > 1$, $E_1 = T_1$, are i.i.d. with exponential law. Since the knowledge of all E_n is equivalent to the knowledge of the process path, we can start with the E_n's and place an error structure on them.

Consider the error structure

$$S = \big(\mathbb{R}_+, \mathcal{B}(\mathbb{R}_+), e^{-x}\,dx, \boldsymbol{d}, u \xrightarrow{\gamma} xu'^2(x)\big),$$

closure of the pre-structure defined on $C_k^\infty(\mathbb{R}_+)$, and define

$$(\Omega, \mathcal{A}, \mathbb{P}, \mathbb{D}, \Gamma) = S^{\otimes \mathbb{N}^*}$$

with the random variables E_n being the coordinate mappings. We have

$$\begin{aligned} \Gamma\big[E_n\big] &= E_n \quad n \geq 1 \\ \Gamma\big[E_m, E_n\big] &= 0 \quad m \neq n. \end{aligned}$$

Lemma. *Setting*

$$D\big[E_n\big] = -1_{]T_{n-1}, T_n]}(t)$$

defines a gradient with value in $\mathcal{H} = L^2(\mathbb{R}_+)$.

Indeed

$$\int_0^\infty 1_{]T_{n-1}, T_n]}(t)\,dt = E_n = \Gamma\big[E_n\big].$$

Among the attractive properties of this gradient is the following.

Proposition VI.32. *Let* $U = \varphi\big(E_1, \ldots, E_n\big)$ *for* $\varphi \in \mathcal{C}^1 \cap \mathrm{Lip}(\mathbb{R}^n)$, *then*

$$U = \mathbb{E}U + \int_0^\infty K_s d\big(N_s - s\big),$$

where K_s *is the predictable projection of the process* $D[U](s)$.

For the proof we refer to Bouleau–Hirsch [1991], Chapter V, Section 5.

Corollary VI.33. *The adjoint operator* δ *coincides with the integral with respect to* $N_t - t$ *on predictable stochastic processes of* $L^2(\mathbb{P}, \mathcal{H})$.

Proof. If H_s is a predictable process in $L^2(\mathbb{P}, \mathcal{H})$, the proposition implies the equality

$$\mathbb{E}\left[U \int_0^\infty H_s\, d\big(N_s - s\big)\right] = \mathbb{E}\left[\int_0^\infty D[U](s) H_s\, ds\right]. \tag{6}$$

It then follows that $H_s \in \operatorname{dom} \delta$ and $\delta[H] = \int_0^\infty H_s d\big(N_s - s\big)$. ◇

Although this error structure yields new integration by parts formulae different from the preceding ones, on very simple random variables it yields the same internalization formula.

Let $X = F(\alpha T_1, \dots, \alpha T_n)$, $F \in \mathcal{C}^1 \cap \mathrm{Lip}$ as before. Then

$$\frac{d}{d\alpha}\mathbb{E}\big[F(\alpha T_1, \dots, \alpha T_n)\big] = \mathbb{E}\Big[\sum_{i=1}^{n} T_i F_i'(\alpha T_1, \dots, \alpha T_n)\Big],$$

whereas

$$D[X] = -\sum_{i=1}^{n} \alpha F_i'(\alpha T_1, \dots, \alpha T_n) 1_{]0,T_i]}(s)$$

such that

$$\begin{aligned} \frac{d}{d\alpha}\mathbb{E}[X] &= -\frac{1}{\alpha}\mathbb{E}\int_0^{T_n} D[X](s)\,ds \\ &= -\frac{1}{\alpha}\mathbb{E}\int_0^{\infty} D[X]1_{]0,T_n]}(s)\,ds. \end{aligned}$$

According to (6) this gives

$$\begin{aligned} &= -\frac{1}{\alpha}\mathbb{E}\left[X\int_{]0,T_n]} d(N_s - s)\right] \\ &= \frac{1}{\alpha}\mathbb{E}\big[X(T_n - n)\big], \end{aligned}$$

which is exactly (5).

Appendix. Comments on current research

Before tackling applications of error calculus in finance and physics, let us indicate some topics of active research.

A useful feature of error structures is to allow Lipschitz calculations: Lipschitz functions operate on the domain of Γ and images by Lipschitz functions are always possible. A natural question is to extend these properties from Lipschitz functions defined on $\mathbb{R}^d$ to Lipschitz functions defined on general metric spaces. This supposes a metric to be available on the basic space Ω of $(\Omega, \mathcal{A}, \mathbb{P}, \mathbb{D}, \Gamma)$. Quite significant progresses have been done in this direction thanks to the *intrinsic distance* defined by M. Biroli and U. Mosco [1991] and [1995], and the study of Lipschitz properties in error structures has been connected (Hirsch [2003]) with more abstract approaches of metrics in measurable spaces (Weaver [1996] and [2000]).

Some aspects of the language of Dirichlet forms for error calculus seem unsatisfactory and require certain improvements. First the fact that only sufficient conditions are at present known for the closedness of a Dirichlet form on $\mathbb{R}^d$, $d > 1$; G. Mokobodzki has given (unpublished) lectures concerning this issue and the conjecture of extending Proposition III.16 to all error structures. An other point to be enlightened would be to construct a practically convenient local definition of the generator domain $\mathcal{D}(A)$.

Eventually, let us mention the question of obtaining an error structure $(\Omega, \mathcal{A}, \mathbb{P}, \mathbb{D}, \Gamma)$ from statistical experiments. The connection of Γ with the *Fisher information matrix* $J(\theta)$ of a parametrized statistical model was sketched in Bouleau [2001]. The connecting relation

$$\underline{\underline{\Gamma}}[I, I^t] = J^{-1}(\theta)$$

is intuitively natural since several authors have noted that the inverse matrix $J^{-1}(\theta)$ describes the accuracy of the knowledge of θ. The relation is also algebraically stable by images and products what provides a special interest to this connection. A thesis (Ch. Chorro) is underway on this topic.

Bibliography for Chapter VI

K. Bichteler, J. B. Gravereau, and J. Jacod, *Malliavin Calculus for Processes with Jumps*, Gordon & Breach, 1987.

M. Biroli and U. Mosco, Formes de Dirichlet et estimations structurelles dans les milieux dicontinus, *C.R. Acad. Sci. Paris Sér. I* **313** (1991), 593–598.

M. Biroli and U. Mosco, A Saint Venant type principle for Dirichlet forms on discontinuous media, *Ann. Math. Pura Appl.* **169** (1995), 125–181.

V. Bogachev and M. Röckner, Mehler formula and capacities for infinite dimensional Ornstein–Uhlenbeck process with general linear drift, *Osaka J. Math.* **32** (1995) 237–274.

N. Bouleau, Construction of Dirichlet structures, in: *Potential Theory - ICPT94*, (Král, Lukeš, Netuka, Veselý, eds.) Walter de Gruyter, 1995.

N. Bouleau, *Processus stochastiques et Applications*, Hermann, 2000.

N. Bouleau, Calcul d'erreur complet Lipschitzien et formes de Dirichlet, *J. Math. Pures Appl.* **80**(9) (2001), 961–976.

N. Bouleau and Ch. Chorro, Error Structures and Parameter Estimation, *C.R. Acad. Sci. Paris Sér. I*, 2003.

N. Bouleau and F. Hirsch, Propriétés d'absolue continuité dans les espaces de Dirichlet et application aux équations differentielles stochastiques, in: *Sém. Probab. XX*, Lecture Notes in Math. 1204, Springer-Verlag, 1986.

N. Bouleau and F. Hirsch, *Dirichlet Forms and Analysis on Wiener Space*, Walter de Gruyter, 1991.

E. Carlen and E. Pardoux, Differential calculus and integration by parts on Poisson space, in: *Stochastics, Algebra and Analysis in Classical and Quantum Dynamics* (S. Albeverio et al., eds.), Kluwer, 1990.

D. Chafaï, in *Sur les inegalites de Sobolev Logarithmiques* (C. Ané, S. Blachére, D. Chafaï, P. Fougères, I. Gentil, F. Malrieu, C. Roberto, G. Scheffer, eds.) *Panor. Synthèses*, Soc. Math. France, 2000.

A. Coquio, Forme de Dirichlet sur l'espace canonique de Poisson et application aux équations différentielles stochastiques, *Ann. Inst. H. Poincaré*, Probab. Stat. **29** (1993), 1–36.

L. Decreusefond, Perturbation analysis and Malliavin calculus, *Ann. Appl. Probab.* **8** (2) (1998), 495–523.

D. Feyel and A. de La Pradelle, Espaces de Sobolev Gaussiens, *Ann. Institut Fourier* **39** (1989), 875–908.

F. Hirsch, Intrisic metrics and Lipschitz functions, *J. Evol. Equations* **3** (2003), 11–25.

N. Ikeda and Sh. Watanabe, *Stochastic Differential Equations and Diffusion Processes*, North-Holland Kodansha, 1981.

Sh. Kusuoka, Analytic functionals of Wiener process and absolute continuity, in: *Functional Analysis in Markov Processes*, Proc. Int. Workshop, Katata, Japan, 1981 and Int. Conf., Kyoto, Japan, 1981, Lecture Notes in Math. 923, Springer-Veralg, 1982, 1–46.

P. Malliavin, *Stochastic Analysis*, Springer-Verlag, 1997.

D. Nualart, *The Malliavin Calculus and Related Topics*, Springer-Verlag, 1995.

D. Nualart and J. Vives, Anticipative calculus for the Poisson process based on the Fock space, in: *Sém. Probab. XXIV*, Lecture Notes in Math. **1426**, Springer-Verlag, 1990, 154–165.

D. Ocone, A guide to stochastic calculus of variations, in: *Proc. Stoch. Anal. and Rel. Topics*, Silivri 1988, (S. Üstünel, ed.), Lecture Notes in Math. 1444, Springer-Verlag, 1990, 1–79.

J. Potthof, White noise approach to Malliavin calculus, *J. Funct. Anal.* **71** (1987), 207–217.

N. Privault, Calcul chaotique et calcul variationnel pour le processus de Poisson, *C.R. Acad. Sci. Paris Ser. I* **316** (1993), 597–600.

A. S. Üstünel and M. Zakai, *Transformation of Measure on Wiener Space*, Springer-Verlag, 2000.

S. Watanabe, *Lectures on Stochastic Differential Equations and Malliavin Calculus*, Tata Institute of Fundamental Research, Springer-Verlag, 1984.

N. Weaver, Lipschitz Algebras and Derivations of Von Neumann Algebras, *J. Funct. Anal.* **139** (1996), 261–300; Lipschitz Algebras and Derivations II. Exterior Differentiation, *J. Funct. Anal.* **178** (2000), 64–112.

L. Wu, Construction de l'opérateur de Malliavin sur l'espace de Poisson, *Sém. Probab. XXI*, Lecture Notes in Math. 1247, Springer-Verlag, 1987, 100–113.

Chapter VII

Application to financial models

As discussed in the preceding chapters, error calculus applies to a wide range of situations.

Certain features make it particularly relevant for financial models.

Contemporary finance mainly uses stochastic models involving *stochastic integrals* and *stochastic differential equations.* These objects cannot a priori be defined path by path and display significant local irregularity for their path dependence. The error structures tool is well adapted to such a framework, as shown in Chapter VI, it easily handles stochastic processes.

In addition, since in finance all quantities are random from some point of view, purely deterministic error calculus may prove to be insufficient. Let us recall the comparative table presented in Chapter V, Section 1.2 regarding the various kinds of error calculi. The fact that error structures also manage the correlation of errors is invaluable.

Sections 1 and 2 are devoted to a new approach to the theory of options pricing. More precisely, they provide a new language for this theory using error structures; it is based upon the concept of the instantaneous error structure of a financial asset. The global notion of "martingale measure" is replaced by the instantaneous notion of *unbiased* quantity.

In Section 3, we will focus on the errors for the Black–Scholes model. We will first examine the sensitivity of the model to a change in either parameters values or the paths of the processes used in the hypotheses. We will then study the consequences of errors due solely to the trader regarding pricing and hedging.

Section 4 extends this study to the case of a diffusion model. We emphasize the fact that error structures allow handling errors on functional coefficients and conclude by illustrating this point through the sensitivity to the local volatility.

1 Instantaneous error structure of a financial asset

Shares and currencies are often represented in financial models by Markov processes: diffusion processes or jump processes constructed from processes with stationary independent increments (Lévy processes). In the case of a diffusion process, a typical asset model is a stochastic differential equation driven by Brownian motion:

$$S_t = S_0 + \int_0^t a(s, S_s)\, dB_s + \int_0^t b(s, S_s)\, ds.$$

If we consider that at time t, the quantity S_t is marred by an error introduced due to the fact that instead of exactly S_t we have S_{t+h}, due to a small unknown waiting period between the decision and the operation, by the meaning of operators Γ and A as explained in Chapter I, it would be natural *a priori* to set

$$\begin{aligned}\Gamma[S_t] &= \lim_{h\to 0} \frac{1}{h}\mathbb{E}\big[(S_{t+h} - S_t)^2 \mid \mathcal{F}_t\big]\\ A[S_t] &= \lim_{h\to 0} \frac{1}{h}\mathbb{E}\big[S_{t+h} - S_t \mid \mathcal{F}_t\big],\end{aligned}$$

i.e., supposing here the functions a and b to be regular,

$$\begin{aligned}\Gamma[S_t] &= \frac{d\langle S, S\rangle_t}{dt} = a^2(t, S_t)\\ A[S_t] &= b(t, S_t).\end{aligned}$$

However the knowledge of the three objects

$$\begin{cases}\text{law of } S_t\\ \Gamma[S_t]\\ A[S_t]\end{cases}$$

is overabundant to determine an error structure. Indeed in an error structure, once two of these objects are known, under suitable regularity assumptions the third follows. The instantaneous germ of a Markov process is too rich to build an error structure. This finding is due to the fact that an error structure is the germ of a *symmetric* Markov process which is generally not the case of the modeled asset.

At present, no error theory is available for generators of general Markov processes and their squared field operators. The theory of non-symmetric Dirichlet forms (Berg–Forst [1973], Ma–Röckner [1992], Dellacherie–Meyer, Chapter XIII) only provides a partial answer to this question since it deals with Markov processes whose drift term is "dominated" by the diffusion term and it does not apply to uniform translation or subordinators.

As a matter of fact, in finance, the drift term is often unknown or subjective and does not occur in several questions. It is thus natural in such cases to only take the law of S_t and $\Gamma[S_t]$ into account, as these two objects define an error structure under usual regularity assumptions.

This leads us to define $\Gamma[S_t]$ by

$$\Gamma[S_t] = \frac{d\langle S, S\rangle_t}{dt}.$$

Let us remark that the formula of the functional calculus in error structures

$$\Gamma\big[F(S_t)\big] = F'^2(S_t)\Gamma[S_t]$$

and the Itô formula

$$d\langle F(S), F(S)\rangle_t = F'^2(S_t)d\langle S, S\rangle_t$$

show that the relation

$$\Gamma[S_t] = \frac{d\langle S, S\rangle_t}{dt},$$

if admitted for S_t, is still true for processes $F(S_t)$ with F difference of convex functions (in this case, $F(S_t)$ is a semimartingale) or for processes $F(S_t)$ with F of class $\mathcal{C}^1$ (such processes possess a well-defined bracket, see Meyer [1976]).

In the multivariate case, if $S_t = \big(S_t^1, \dots, S_t^d\big)$ satisfies

$$S_t^i = S_0^i + \int_0^t \sum_{k=1}^{\ell} a_{ik}(s, S_s)\, dB_s^k + \int_0^t b_i(s, S_s)\, ds,$$

where $B = \big(B^1, \dots, B^\ell\big)$ is a standard Brownian motion with values in $\mathbb{R}^\ell$, this approach leads to setting

$$\underline{\underline{\Gamma}}[S_t, S_t^t] = \underline{\underline{a}}(t, S_t)\big(\underline{\underline{a}}(t, S_t)\big)^t.$$

Which condition must satisfy the law of S_t and the matrix $\underline{\underline{\alpha}}(x) = \underline{\underline{a}}(t, x)\big(\underline{\underline{a}}(t, x)\big)^t$ such that the pre-structure

$$\big(\mathbb{R}^d, \mathcal{B}(\mathbb{R}^d), \mu, \mathcal{C}_K^\infty(\mathbb{R}^d), \Gamma\big)$$

where

$$\Gamma[u](x) = \sum_{i,j} \alpha_{ij}(x) \frac{\partial u}{\partial x_i} \frac{\partial u}{\partial x_j}$$

be closable? Only sufficient conditions have been brought out up until now, see Chapter III, Example III.23 (see also Fukushima, Oshima, Takeda [1994], Chapter 3, p. 100 et seq.).

We will first take some simple examples and then draw up an instantaneous error structure from non-necessarily Markov stationary processes.

1.1 Geometrical Brownian motion and homogeneous log-normal error structure. Starting with the famous model of Fisher Black and Myron Scholes

$$dS_t = S_t\big(\sigma\, dB_t + r\, dt\big),$$

which admits the explicit solution

$$S_t = S_0 \exp\left(\sigma B_t + \left(r - \frac{\sigma^2}{2}\right) t\right)$$

we must consider the error structure

$$\Sigma = \big(\mathbb{R}_+, \mathcal{B}(\mathbb{R}_+), \nu, \boldsymbol{d}, \gamma\big)$$

where ν is a log-normal law, image of the normal law $\mathcal{N}(0,t)$ by the map

$$x \to S_0 \exp\left(\sigma x + \left(r - \frac{\sigma^2}{2}\right) t\right)$$

and where

$$\gamma[u](x) = \sigma^2 x^2 u'^2(x)$$

for regular u's (this is the expression of the above principle $\Gamma[S_t] = \frac{d\langle S,S\rangle_t}{dt} = \sigma^2 S_t^2$).

This structure possesses several interesting properties.

a) It is the image of the one-dimensional Ornstein–Uhlenbeck structure:

$$\Sigma_{\mathrm{OU}} = \big(\mathbb{R}, \mathcal{B}(\mathbb{R}), \mathcal{N}(0,t), H^1(\mathcal{N}(0,t)), \Gamma_1 \colon v \to v'^2\big)$$

by means of the application $x \to S_0 \exp\left(\sigma x + \left(r - \frac{\sigma^2}{2}\right) t\right)$. Indeed, let Γ_Σ be the quadratic error operator of the image, by setting $\xi(x) = S_0 \exp\big(\sigma x + (r - \frac{\sigma^2}{2})t\big)$, we obtain

$$\Gamma_\Sigma[u](y) = \mathbb{E}\big[\Gamma_1[u \circ \xi] \mid \xi = y\big] = u'^2(y)\sigma^2 y^2.$$

b) It then follows that the domain of γ is explicitly

$$\boldsymbol{d} = \big\{u \in L^2(\nu) \colon x \to x^2 u'^2(x) \in L^1(\nu)\big\}.$$

c) As image of the structure Σ_{OU}, it also satisfies the Poincaré inequality:

$$\mathrm{var}_\nu[u] \le \mathbb{E}_\nu\big[\gamma[u]\big].$$

d) It yields a *homogeneous* quadratic error

$$\gamma[I] = \sigma^2 (I)^2,$$

where I is the identity map $x \to x$. In other words, it represents a *constant proportional error* from the standpoint of a physicist: if S_t is modeled by this structure, then

$$\frac{\mathbb{E}\big[\Gamma[S_t] \mid S_t = x\big]}{x^2} = \frac{\mathbb{E}\big[\Gamma[S_t] \mid S_t = 2x\big]}{(2x)^2} = \sigma^2.$$

e) Denoting $(A, \mathcal{D}A)$ the generator of this structure Σ, we have $I \in \mathcal{D}A$ and

$$A[I](y) = y \frac{1}{2}\left[\frac{\sigma^2}{2} + r - \frac{1}{t} \log \frac{y}{S_0}\right]$$

in particular $A[I](y_0) = 0$ if

$$y_0 = S_0 \exp\left\{\left(r + \frac{\sigma^2}{2}\right) t\right\} = \|S_t\|_{L^2}.$$

Exercise (A three-currency model). Let $\left(B_t^1, B_t^2\right)$ be a standard Brownian motion with values in $\mathbb{R}^2$, the diffusion processes

$$(*)\qquad \begin{aligned} dS_t^{ij} = S_t^{ij}\big\{&\left(a_1^j(t) - a_1^i(t)\right) dB_t^1 + \left(a_2^j(t) - a_2^i(t)\right) dB_t^2 \\ &+ \tfrac{1}{2}\left[(a_1^j(t) - a_1^i(t))^2 + (a_2^j(t) - a_2^i(t))^2 + b^j(t) - b^i(t)\right]dt\big\}\end{aligned}$$

for $i, j = 1, 2, 3$, where the functions a_1^i, a_2^i, b^i are, let us say bounded continuous and deterministic, identically satisfy

$$S_t^{ij} = S_t^{ik} S_t^{kj}, \quad 1 \le i, j, k \le 3$$

once these relations have been fulfilled at $t = 0$, as it can be verified using Itô calculus.

For example with $b^i(t) = 0$, $a_1^i(t) = \cos[(i-1)2\pi/3]$ and $a_2^i(t) = \sin[(i-1)2\pi/3]$ for $i = 1, 2, 3$, we obtain the model

$$\begin{cases} dS_t^{12} = S_t^{12}\left(-\frac{3}{2}dB_t^1 + \frac{\sqrt{3}}{2}dB_t^2 + \frac{3}{2}dt\right) & dS_t^{21} = S_t^{21}\left(\frac{3}{2}dB_t^1 - \frac{\sqrt{3}}{2}dB_t^2 + \frac{3}{2}dt\right) \\ dS_t^{23} = S_t^{23}\left(-\sqrt{3}dB_t^2 + \frac{3}{2}dt\right) & dS_t^{32} = S_t^{32}\left(\sqrt{3}dB_t^2 + \frac{3}{2}dt\right) \\ dS_t^{31} = S_t^{31}\left(\frac{3}{2}dB_t^1 + \frac{\sqrt{3}}{2}dB_t^2 + \frac{3}{2}dt\right) & dS_t^{13} = S_t^{13}\left(-\frac{3}{2}dB_t^1 - \frac{\sqrt{3}}{2}dB_t^2 + \frac{3}{2}dt\right) \end{cases}$$

where the three currencies play a symmetric role and the six rates S_t^{ij}, $i \ne j$ are all submartingales with the same law (up to the multiplication by their initial values) which is the law of the process $\exp(\sqrt{3}B_t)$.

What does the preceding method provide as instantaneous error structure from a model like $(*)$?

We observe that

$$\frac{d\langle S^{ij}, S^{ij}\rangle_t}{dt} = \left(\left(a_1^j(t) - a_1^i(t)\right)^2 + \left(a_2^j(t) - a_2^i(t)\right)^2\right)(S_t^{ij})^2.$$

Thus we must have

$$(**)\qquad \Gamma[S^{ij}] = \left((a_1^j - a_1^i)^2 + (a_2^j - a_2^i)^2\right)(S^{ij})^2.$$

The method leads to consider three positive quantities

$$S^{ij} = \exp\left\{\left(a_1^j - a_1^i\right)X + \left(a_2^j - a_2^i\right)Y + b_j - b_i\right\}$$

for $i, j = 1, 2, 3$, where the numbers a_1^i, a_2^i, b^i are constants and where X, Y are the coordinate mappings defined on the error structure

$$(\Omega, \mathcal{A}, \mathbb{P}, \mathbb{D}, \Gamma) = \left(\mathbb{R}^2, \mathcal{B}(\mathbb{R}^2), \mathcal{N}_2(0, I), H^1(\mathcal{N}_2(O, I)), |\nabla|^2\right).$$

Then $(**)$ is satisfied. The above structure $(\Omega, \mathcal{A}, \mathbb{P}, \mathbb{D}, \Gamma)$ is the Ornstein–Uhlenbeck structure in dimension two. Its generator satisfies

$$A[u](x, y) = \frac{1}{2}\left(\frac{\partial^2 u}{\partial x^2} + \frac{\partial^2 u}{\partial y^2}) - \frac{1}{2}(x\frac{\partial u}{\partial x} + y\frac{\partial u}{\partial y}\right)$$

hence, since X, Y are the coordinate maps

$$A[X] = -\frac{1}{2}X$$

and

$$A[Y] = -\frac{1}{2}Y$$

and by the general formula

$$\begin{aligned} A[F(X,Y)] = F_1'(X,Y)A[X] + F_2'(X,Y)A[Y] \\ + \frac{1}{2}(F_{11}''(X,Y)\Gamma[X] + F_{22}''(X,Y)\Gamma[Y]) \end{aligned}$$

we have

$$\begin{aligned} A[S^{ij}] = S^{ij}\Big[(a_1^j - a_1^i)\Big(-\frac{1}{2}X\Big) + (a_2^j - a_2^i)\Big(-\frac{1}{2}Y\Big) \\ + \frac{1}{2}(a_1^j - a_1^i)^2 + \frac{1}{2}(a_2^j - a_2^i)^2\Big]. \end{aligned}$$

Thus, this structure satisfies

$$A[S^{ij}] = \frac{S^{ij}}{2}\Big[(a_1^j - a_1^i)^2 + (a_2^j - a_2^i)^2 - \log S^{ij} + b_j - b_i\Big]$$

and

$$\frac{A[S^{ij}]}{S^{ij}} + \frac{A[S^{ji}]}{S^{ji}} = \frac{\Gamma[S^{ij}]}{(S^{ij})^2} = \frac{\Gamma[S^{ji}]}{(S^{ji})^2}. \qquad \diamond$$

1.2 Instantaneous structure associated with a stationary process. Under the same set of hypotheses, it is possible to define an error structure tangent at time $t = 0$ to a strictly stationary process.

Proposition VII.1. *Let $(X_t)_{t\in\mathbb{R}}$ be a strictly stationary process with values in $\mathbb{R}^d$ and with continuous sample paths. Let ν denote the law of X_0. We assume that $\forall f \in \mathcal{C}_K^\infty(\mathbb{R}^d)$, the limit*

$$\lim_{t\to 0}\frac{1}{2t}\mathbb{E}\big[f(X_{-t}) - 2f(X_0) + f(X_t) \mid X_0 = x\big]$$

exists in $L^2(\mathbb{R}^d, \nu)$. Let $A[f]$ be this limit and $\Gamma[f]$ be defined via

$$\Gamma[f] = A[f^2] - 2fA[f],$$

then $(\mathbb{R}^d, \mathcal{B}(\mathbb{R}^d), \nu, \mathcal{C}_K^\infty(\mathbb{R}^d), \Gamma)$ is a closable error pre-structure.

Proof. a) The operator A defined on $\mathcal{C}_K^\infty$ is symmetric in $L^2(\nu)$. Indeed, let $f, g \in \mathcal{C}_K^\infty(\mathbb{R}^d)$,

$$\langle A[f], g\rangle_{L^2(\nu)} = \lim_{t\to 0} \frac{1}{2t}\mathbb{E}\big[\big(f(X_{-t}) - 2f(X_0) + f(X_t)\big)g(X_0)\big],$$

which, by stationarity, is equal to

$$\begin{aligned} &= \lim_{t\to 0} \frac{1}{2t}\mathbb{E}\big[f(X_0)g(X_t) - 2f(X_0)g(X_0) + f(X_0)g(X_{-t})\big] \\ &= \langle f, A[g]\rangle. \end{aligned}$$

b) The operator A is negative. Let $f \in \mathcal{C}_K^\infty(\mathbb{R}^d)$,

$$\begin{aligned} \langle A[f], f\rangle_{L^2(\nu)} &= \lim_{t\to 0} \frac{1}{2t}\mathbb{E}\big[\big(f(X_{-t}) - 2f(X_0) + f(X_t)\big)f(X_0)\big] \\ &= \lim_{t\to 0} \frac{1}{t}\mathbb{E}\big[\big(f(X_t) - f(X_0)\big)f(X_0)\big] \\ &= \lim_{t\to 0} \frac{1}{t}\mathbb{E}\big[f(X_t)\big(f(X_0) - f(X_t)\big)\big]. \end{aligned}$$

Taking the half sum of the last two results yields

$$= \lim_{t\to 0} \frac{-1}{2t}\mathbb{E}\big[\big(f(X_t) - f(X_0)\big)^2\big].$$

c) Let us remark that if $f \in \mathcal{C}_K^\infty(\mathbb{R}^d)$

$$\begin{aligned} \Gamma[f](x) &= \lim_{t\to 0} \frac{1}{2t}\mathbb{E}\big[f^2(X_{-t}) - 2f^2(X_0) + f^2(X_t) \\ &\qquad - 2f(X_0)\big(f(X_{-t}) - 2f(X_0) + f(X_t)\big) \mid X_0 = x\big] \\ &= \lim_{t\to 0} \frac{1}{2t}\mathbb{E}\big[\big(f(X_{-t}) - f(X_0)\big)^2 + \big(f(X_t) - f(X_0)\big)^2 \mid X_0 = x\big]. \end{aligned}$$

Using the continuity of the paths, it then follows that Γ satisfies the functional calculus of order $\mathcal{C}^\infty \cap \text{Lip}$. Hence, the term written in the statement of the proposition is an error pre-structure and is closable by Lemma III.24. ◇

2 From an instantaneous error structure to a pricing model

If we consider that at the present time, a given financial quantity (the price of an asset) is erroneous due to the turbulence of the rates, then one of the simplest models consists of assuming the quantity to be a random variable defined on an error structure

$(\Omega, \mathcal{A}, \mathbb{P}, \mathbb{D}, \Gamma)$, such that $\Gamma[S] = \sigma^2 S^2$ in other words, using the image structure by S,

(1) $$\Gamma_S[I](x) = \sigma^2 x^2$$

with the observed spot value s_0 being such that

(2) $$A_S[I](s_0) = 0.$$

Indeed, equation (1) regarding the variance of the error expresses the restlessness of prices while equation (2) concerning the bias, expresses the fact that in s_0 we are not aware whether the rate is increasing or decreasing; hence the bias, i.e. the instantaneous skew, must vanish at s_0.

Then, by the change of variable formula for regular F

$$A_S[F] = F' A_S[I] + \frac{1}{2} F'' \Gamma_S[I],$$

we obtain

(3) $$A_S[F](s_0) = \frac{1}{2} F''(s_0)\sigma^2 s_0^2.$$

In other words, if F is non-affine, *there is bias on* $F(S)$. That means that the right price to be ascribed to the quantity $F(S)$ is not $F(s_0)$, but rather

(4) $$F(s_0) + \frac{1}{2} F''(s_0)\sigma^2 s_0^2 h.$$

We have translated the Black–Scholes–Merton method of pricing an option into the language of error structures by reducing the time interval between the present time and the exercise time to an infinitely small h, with the payoff of the option being in this instance $F(S)$.

2.1 From instantaneous Black–Scholes to Black–Scholes. a) Let us now consider a time interval $[0, T]$ and a European option of payoff $F(S)$ at time T.

Let us share $[0, T]$ into n subintervals of length $h = \frac{T}{n}$ i.e. $\left[\frac{kT}{n}, \frac{(k+1)T}{n}\right]$, $k = 0, \ldots, n-1$. By formula (4), the value at time t of a quantity whose value is $F_{t+h}(S)$ at time $t + h$ is

(5) $$F_t(S) = F_{t+h}(S) + \frac{1}{2} \frac{d^2 F_{t+h}}{dx^2}(S)\sigma^2 S^2 h.$$

Introducing the operator B

(6) $$B[u](x) = \frac{1}{2}\sigma^2 x^2 u''(x),$$

which retains only the second-order term of the generator A_S of the error structure of S, relation (5) can be written as

$$F_t = F_{t+h} + hBF_{t+h}. \tag{7}$$

Let us now transform this relation in order to apply convenient hypotheses of the theory of operators semigroups: Relation (7) can be written

$$F_{t+h} = F_t - hBF_{t+h} = F_t - hBF_t + h^2B^2F_{t+h},$$

thus, neglecting the terms in h^2,

$$F_t = (I - hB)^{-1}F_{t+h}.$$

The induction now yields

$$F_0 = \left(I - \frac{T}{n}B\right)^{-n} F \tag{8}$$

and the following lemma indicates that F_0 converges to

$$Q_T F,$$

where $(Q_t)_{t\geq 0}$ is the semigroup of generator B.

Lemma VII.2. *Let B be the generator of a strongly continuous contraction semigroup Q_t on a Banach space $\mathcal{B}$. Then for all $x \in \mathcal{B}$*

$$\left(I - \frac{t}{n}B\right)^{-n} x \xrightarrow[n\uparrow\infty]{} Q_t x \quad \text{in } \mathcal{B}.$$

Proof. Let us introduce the so-called resolvent family of operators R_λ defined by

$$R_\lambda x = \int_0^\infty e^{-\lambda t} Q_t x \, dt, \quad \lambda > 0.$$

It can easily be proved that R_λ is a bounded operator, such that $(\lambda I - B)R_\lambda x = x$, $\forall x \in \mathcal{B}$, and $R_\lambda(\lambda I - B)x = x$, $\forall x \in \mathcal{D}B$. We also have $\forall x \in \mathcal{B}$, the relation

$$\lambda^n(\lambda I - B)^{-n}x = (\lambda R_\lambda)^n x = \frac{\lambda^n}{(n-1)!}\int_0^\infty e^{-\lambda s}s^{n-1}Q_s x \, ds.$$

Hence,

$$\left(I - \frac{t}{n}B\right)^{-n} x = \left(\frac{n}{t}R_{\frac{n}{t}}\right)^n x = \frac{n^n}{(n-1)!}\int_0^\infty e^{-nu}u^{n-1}Q_{tu}x \, du.$$

The result is now derived from the fact that the probability measure

$$\frac{n^n}{(n-1)!}e^{-nu}u^{n-1}1_{[0,\infty[}(u)\,du$$

narrowly converges to the Dirac measure at point 1 as n tends to infinity and that the map $u \to Q_{tu}x$ is bounded and continuous from $\mathbb{R}_+$ into $\mathcal{B}$. ◇

The semigroup Q_t obtained is the modified heat semigroup:

$$(Q_tF)(x) = \int_{\mathbb{R}} F\left(xe^{\sigma y - \frac{\sigma^2}{2}t}\right)\frac{1}{\sqrt{2\pi t}}e^{-\frac{y^2}{2t}}\,dy.$$

We finally obtain the pricing formula

(9) $$F_0(S) = \big(Q_TF\big)(S).$$

b) In the preceding argument, the interest rate r to be taken into account in the reasoning, was actually omitted.

We must return to formulae (1) and (2) and modify them into

(1bis) $$\Gamma_S[I](x) = \sigma^2x^2$$

(2bis) $$A_S[I](s_0) = rs_0.$$

Formula (2bis) expresses the fact that at the present time where $S = s_0$, the proportional instantaneous skew displays value r.

Formula (3) then becomes

(3bis) $$A_S[F](s_0) = rs_0F'(s_0) + \frac{1}{2}\sigma^2s_0F''(s_0),$$

hence the value to be ascribed the quantity $F(S)$ is

(4bis) $$e^{-rh}\left\{F(s_0) + h\left[rs_0F'(s_0) + \frac{1}{2}\sigma^2s_0^2F''(s_0)\right]\right\}.$$

This change modifies formula (5) in the following manner:

(5bis) $$F_t(S) = e^{-rh}\left\{F_{t+h}(S) + h\left[rS\frac{dF_{t+h}}{dx}(S) + \frac{1}{2}\sigma^2S^2\frac{d^2F_{t+h}}{dx^2}(S)\right]\right\}.$$

Neglecting the terms in h^2 yields

$$F_t = (I - hB)^{-1}F_{t+h}$$

with

(6bis) $$B[u](x) = \frac{1}{2}\sigma^2x^2u''(x) + rxu'(x) - ru(x).$$

Using this notation the remainder of the argument stays unchanged and we have, as before,

$$\text{(8bis)} \qquad F_0 = \left(I - \frac{T}{n}B\right)^{-n} F$$

and as n tends to infinity

$$F_0 = \lim_n \left(I - \frac{T}{n}B\right)^{-n} F = e^{TB}F = P_T F,$$

where $(P_t)_{t\geq 0}$ is the semigroup with generator B, which now becomes

$$\text{(10)} \qquad (P_t F)(x) = e^{-rt} \int_{\mathbb{R}} F\left(xe^{\sigma y + \left(r - \frac{\sigma^2}{2}\right)t}\right) \frac{1}{\sqrt{2\pi t}} e^{-\frac{y^2}{2t}}\, dy.$$

We obtain the Black–Scholes formula: the value of the option at time t is given by the function $P_{T-t}F$ taken on the price of the asset S at time t.

c) Let us return to the case $r = 0$ in order to examine the question of hedging.

Formula (5) indicates that the function to take on the spot of the asset price for obtaining the value of the option at time t, i.e. F_t, satisfies

$$F_t(x) = F_{t+h}(x) + \frac{1}{2}\frac{\partial^2 F_{t+h}}{\partial x^2}(x)\sigma^2 x^2 h,$$

thus

$$\text{(11)} \qquad \frac{\partial F_t}{\partial t}(x) + \frac{1}{2}\sigma^2 x^2 \frac{\partial^2 F_t}{\partial x^2}(x) = 0 \quad \forall x \in \mathbb{R}_+,\ \forall t \in [0, T].$$

Instantaneous hedging, i.e. what is called in the usual language of practioners the question of *delta neutral hedging portfolio*, may be tackled without additional assumptions.

Let k be the quantity of assets that we must possess at time t in order that the portfolio consisting of the sold option (the value of the option with the sign minus) and this quantity of assets be insensitive to errors on S_t?

We have to write that the variance of the error on $kS_t - F_t(S_t)$ is zero on the spot value s_t. This is

$$\Gamma_S[kI - F_t](s_t) = 0$$

or by the functional calculus

$$(k - \frac{\partial F_t}{\partial x}(s_t))^2 \Gamma_S[I](s_t) = 0,$$

i.e.,

$$k = \frac{\partial F_t}{\partial x}(s_t).$$

We find, not surprisingly, the usual hedging portfolio.

Now to study the hedging from time 0 to time T, we need hypotheses on the stochastic process S_t.

Assuming the asset price S_t to be a continuous semi-martingale, the Itô formula gives

$$F_t(S_t) = F_0(S_0) + \int_0^t \frac{\partial F_s}{\partial x}(S_s) dS_s + \int_0^t \frac{\partial F_s}{\partial s}(S_s)\, ds + \frac{1}{2}\int_0^t \frac{\partial^2 F_s}{\partial x^2}(S_s) d\langle S, S\rangle_s.$$

observe, once the following condition is fulfilled

$$d\langle S, S\rangle_s = \sigma^2 S_s^2\, ds$$

from (11)

$$F_t(S_t) = F_0(S_0) + \int_0^t \frac{\partial F_s}{\partial x}(S_s)\, dS_s, \tag{12}$$

we conclude that the *value of the option* $F_t(S_t)$ *can be written as a constant plus a stochastic integral with respect to* S_t. In other words exact hedging is occurring.

In short, this approach shows that

(i) the value of the European option with payoff $F(S_T)$ at time T is $\big(P_T F\big)(S_0)$ at time 0,

(ii) once $F_0 = P_T F$ has been calculated, a hedging portfolio exists consisting of $\frac{\partial F_t}{\partial x}(S_t)$ assets at time t by means of formula (12).

We note that if S_t satisfies only an inequality $d\langle S, S\rangle_t \le \sigma^2 S_t^2\, dt$ [resp. $d\langle S, S\rangle_t \ge \sigma^2 S_t^2\, dt$], then when F is a convex function (which implies that $F_t = P_{T-t} F$ is convex as well by (10))

$$F_t(S_t) \le F_0(S_0) + \int_0^t \frac{\partial F_s}{\partial x}(S_s)\, dS_s$$

and the above portfolio (ii) is a hedging strategy with benefit [resp. deficit]. This conclusion is inverted when F is concave.

We can also remark that if S_t is a semi-martingale with jumps whose continuous martingale part satisfies $d\langle S^c, S^c\rangle_t = \sigma^2 S_t^2\, dt$, then the Itô formula, taking (11) into account, yields

$$F_t(S_t) = F_0(S_0) + \int_0^t \frac{\partial F_s}{\partial x}(S_{s-})\, dS_s + \sum_{s \le t}\left[F_s(S_s) - F_s(S_{s-}) - \frac{\partial F_s}{\partial x}(S_{s-})\, \Delta S_s\right],$$

and we conclude that the portfolio with $\frac{\partial F_s}{\partial x}(S_{s-})$ assets at time s yields a deficit if F is convex, regardless of the jumps: positive, negative or mixed.

d) When $r \neq 0$, the conclusions are similar.

Equation (11) becomes

$$\frac{\partial F_t}{\partial t} + \frac{1}{2}\sigma^2 x^2 \frac{\partial^2 F_t}{\partial x^2} + rx\frac{\partial F_t}{\partial x} - rF_t = 0 \tag{13}$$

and if the price of the asset is a continuous semi-martingale such that $d\langle S, S\rangle_t = \sigma^2 S_t^2$, the Itô formula and (13) give

$$e^{-rt} F_t(S_t) = F_0(S_0) + \int_0^t \frac{\partial F_s}{\partial x}(S_s)\, d\big(e^{-rs} S_s\big) \tag{14}$$

and the conclusions are analogous.

2.2 From an instantaneous error structure to a diffusion model. The preceding approach naturally extends to more general hypotheses.

Let us suppose that the instantaneous error structure is defined for the vector asset $S = \big(S^1, S^2, \dots, S^d\big)$ on, for example, the error structure $(\Omega, \mathcal{A}, \mathbb{P}, \mathbb{D}, \Gamma)$, by $\underline{\underline{\Gamma}}[S, S^t] = \big(\alpha_{ij}(S)\big)_{ij}$, i.e. using the image space through S

$$\underline{\underline{\Gamma}}_S[I, I^t](x) = \big(\alpha_{ij}(x)\big)_{ij} \quad \forall x \in \mathbb{R}^d \tag{15}$$

where $\big(\alpha_{ij}(x)\big)$ is a positive symmetric matrix, or

$$\Gamma_S[X^i, X^j](x) = \alpha_{ij}(x)$$

denoting $I = (X^1, \dots, X^d)$ the identity map from $\mathbb{R}^d$ onto $\mathbb{R}^d$.

If we start with the hypothesis

$$A_S[X^i](s_0) = 0 \quad \forall i = 1, \dots, d \tag{16}$$

which expresses the lack of any skew on the spot price $s_0 = \big(s_0^1, \dots, s_0^d\big)$, we then have for regular F from $\mathbb{R}^d$ into $\mathbb{R}$

$$A_S[F] = \sum_i F_i' A_S[X^i] + \frac{1}{2} \sum_{ij} F_{ij}'' \Gamma_S[X^i, X^j]$$

hence

$$A_S[F](s_0) = \frac{1}{2} \sum_{ij} \alpha_{ij}(s_0) F_{ij}''(s_0) \tag{17}$$

and the value to be ascribed the quantity $F(S)$ at the present time is not $F(s_0)$, but rather

$$F(s_0) + A_S[F](s_0)h = F(s_0) + \frac{h}{2} \sum_{ij} \alpha_{ij}(s_0) F_{ij}''(s_0), \tag{18}$$

where h expresses the size of the errors which is the same order of magnitude both for variances and biases (see Chapter I).

a) The incremental reasoning on the interval $[0, T]$ shared in sub-intervals $\left(\frac{mT}{n}, \frac{(m+1)T}{n}\right)$ for the pricing of a European option with exercise time T and pay-off $F\big(S_T\big)$ can then proceed as follows.

If the value of the option at time t is expressed by the function F_t on the price of the asset, by (18) we must have

$$F_t = F_{t+h} + \frac{h}{2}\sum_{ij}\alpha_{ij}\frac{\partial^2 F_{t+h}}{\partial x_i \partial x_j} \tag{19}$$

or

$$F_t = (I + hB)F_{t+h} \tag{20}$$

where B is the operator

$$B[u](x) = \frac{1}{2}\sum_{ij}\alpha_{ij}(x)\frac{\partial^2 u}{\partial x_i \partial x_j}$$

and $h = \frac{T}{n}$.

Neglecting the terms in h^2 allows rewriting equation (20):

$$F_t = \left(I - \frac{T}{n}B\right)^{-1} F_{t+h}$$

and the induction yields

$$F_0 = \left(I - \frac{T}{n}B\right)^{-n} F.$$

Supposing coefficients α_{ij} such that B is the generator of a strongly continuous semigroup in order we may apply Lemma VII.2, we obtain at the limit:

$$F_0 = \lim_{n\to\infty}\left(I - \frac{T}{n}B\right)^{-n} F = e^{TB} = Q_T F \tag{21}$$

where $\big(Q_t\big)_{t\geq 0}$ is the semigroup with generator B.

The function $F_t \colon \mathbb{R}^d \to \mathbb{R}$ thus satisfies $F_t = Q_{T-t}F$ or in other terms

$$\begin{cases} \dfrac{\partial F_t}{\partial t} + \dfrac{1}{2}\displaystyle\sum_{ij}\alpha_{ij}\frac{\partial^2 F_t}{\partial x_i \partial x_j} = 0 \\ \\ F_T = F. \end{cases} \tag{22}$$

b) If the interest rate is no longer zero, a discount factor $\exp\big(-\int_0^t r(s)\,ds\big)$ between 0 and t is present. Equation (16) must then be changed into

$$A_S[X^i](s_0) = r_0 s_0^i, \quad i = 1, \ldots, d,$$

and equation (19) becomes

$$F_t = e^{-r(t)h}\left[F_{t+h} + \frac{h}{2}\sum_{ij}\alpha_{ij}\frac{\partial^2 F_{t+h}}{\partial x_i \partial x_j} + hr(x)\overrightarrow{X}\cdot\overrightarrow{\nabla F}_{t+h}\right],$$

where $\overrightarrow{X} = (x_1, \dots, x_d)$ and $\overrightarrow{\nabla F}_{t+h} = \left(\frac{\partial F_{t+h}}{\partial x_i}\right)_{i=1,\dots,d}$.

The price of the option is thus given by the function F_0 taken on the spot of the price with

$$F_0 = \lim_{n\uparrow\infty}\prod_{k=1}^{n}\left(I + \frac{T}{n}B_k\right)F,$$

where

$$B_k[u](x) = \frac{1}{2}\sum_{ij}\alpha_{ij}(x)\frac{\partial^2 u}{\partial x_i \partial x_j} + r\left(\frac{kT}{n}\right)\sum_i x_i\frac{\partial u}{\partial x_i} - r\left(\frac{kT}{n}\right)u.$$

It follows that the function F_t to be taken on the spot for obtaining the discounted value of the option at time t satisfies

$$(23)\qquad \begin{cases} \dfrac{\partial F_t}{\partial t} + \dfrac{1}{2}\displaystyle\sum_{ij}\alpha_{ij}(x)\frac{\partial^2 F_t}{\partial x_i \partial x_j} + r(t)\sum_i x_i\frac{\partial F_t}{\partial x_i} - r(t)F_t = 0 \\ \\ F_T = F. \end{cases}$$

c) Regarding the hedging, let us return to the simplest case where $r = 0$.

As in the Black–Scholes case, the instantaneous hedging, i.e. the question of *finding a delta neutral portfolio*, may be answered without additional assumption.

A portfolio consisting of k_i assets S^i $(i = 1, \dots, d)$ and the sold option is *insensitive to errors on* S_t if the variance of the error on $\sum_i k_i S_t^i - F_t(S_t)$ vanishes on the spot value $s_t = (s_t^1, \dots, s_t^d)$. With our notation the condition is

$$\underset{=}{\Gamma_S}\left[\sum_i k_i X^i - F_t(I)\right](s_t) = 0$$

or with the help of the functional calculus

$$(\vec{k} - \vec{\nabla F_t}(s_t))^t \underset{=}{\Gamma_S}[I, I^t](s_t)(\vec{k} - \vec{\nabla F_t}(s_t)) = 0$$

with $\vec{k} = (k_1, \dots, k_d)^t$.

Thus, as soon as the functions α_{ii} do not vanish, necessarily

$$k_i = \frac{\partial F_t}{\partial x_i}(s_t), \quad i = 1, \dots, d.$$

which is the expected result.

Now for the hedging from time 0 to time T, let us suppose that the price of the asset is modeled by a continuous semi-martingale S_t such that

$$\frac{d\langle S^i, S^j\rangle_t}{dt} = \alpha_{ij}(S_t), \tag{24}$$

then equation (22) and the Itô formula show that

$$F_t(S_t) = F_0(S_0) + \sum_i \int_0^t \frac{\partial F}{\partial x_i}(S_s)\, dS_s. \tag{25}$$

After the pricing $F_0(S_0)$, the portfolio consisting of $\frac{\partial F}{\partial x_i}(S_s)$ assets S^i is therefore an exact hedging of the option.

Exercise. The instantaneous error structure taken at the beginning of the construction may involve an error on the volatility. Then we have to connect with the numerous published works on "stochastic volatility".

Suppose that the instantaneous error structure is such that

$$\Gamma[S] = S^2\Sigma^2,$$

where Σ is a random variable defined on the same error structure $(\Omega, \mathcal{A}, \mathbb{P}, \mathbb{D}, \Gamma)$ as S, such that Σ also possesses an error.

a) If we suppose

$$\begin{pmatrix} \Gamma[S] & \Gamma[S,\Sigma] \\ \Gamma[S,\Sigma] & \Gamma[\Sigma] \end{pmatrix} = \begin{pmatrix} S^2\Sigma^2 & S\Sigma^2\gamma\rho \\ S\Sigma^2\gamma\rho & \Sigma^2\gamma^2 \end{pmatrix}$$

where γ and ρ are constants, we can translate into terms of instantaneous error structure a model such as the following:

$$\begin{cases} d\delta_t = \delta_t\big(\sigma_t\, dB_t + \mu\, dt\big) \\ d\sigma_t = \sigma_t\big(\gamma\, dW_t + \alpha\, dt\big) \end{cases} \tag{26}$$

where (B_t) and (W_t) are two Brownian motions with correlation ρ (model studied by Hull and White (1987) and by Wiggins (1987)) along with a model like

$$\begin{cases} d\delta_t = \delta_t\big(\sigma_t\, dB_t + \mu\, dt\big) \\ d\sigma_t = \sigma_t\big(\gamma\, dW_t + (\alpha - \beta\sigma_t)\, dt\big) \end{cases} \tag{27}$$

(studied by Scott (1987)).

b) If we suppose

$$\begin{pmatrix} \Gamma[S] & \Gamma[\rho,\Sigma] \\ \Gamma[S,\Sigma] & \Gamma[\Sigma] \end{pmatrix} = \begin{pmatrix} S^2\Sigma^2 & \rho\Sigma\gamma\rho \\ S\Sigma\gamma\rho & \gamma^2 \end{pmatrix}$$

we can express the model

$$\begin{cases} d\delta_t = \delta_t\big(\sigma_t\, dB_t + \mu\, dt\big) \\ d\sigma_t = \gamma\, dW_t + \beta(\alpha - \sigma_t)\, dt \end{cases}$$

studied by Scott (1987) and Stein and Stein (1991) along with the model

$$\begin{cases} d\delta_t = S_t\big(\sigma_t\, dB_t + \mu\, dt\big) \\ d\sigma_t = \gamma\, dW_t + \big(\frac{\delta}{\sigma_t} - \beta\sigma_t\big)\, dt \end{cases}$$

studied by Hull and White (1988) and Heston (1993).

c) If we suppose that Σ can be written $\Sigma = \varphi(Y)$

$$\begin{pmatrix} \Gamma[S] & \Gamma[S,\Sigma] \\ \Gamma[S,\Sigma] & \Gamma[\Sigma] \end{pmatrix} = \begin{pmatrix} S^2\varphi^2(Y) & S\varphi(Y)\gamma\rho \\ S\varphi(Y)\gamma\rho & \gamma^2 \end{pmatrix}$$

we express a model of the type

$$\begin{cases} d\delta_t = \delta_t\big(\varphi(Y_t)\, dB_t + \mu\, dt\big) \\ dY_t = \gamma\, dW_t + \alpha(m - Y_t)\, dt \end{cases}$$

studied by Fouque and Tullie (2001).

In these cases, the reasoning displayed in Sections 2.1 and 2.2, in omitting the discounting for the sake of simplicity, begins as follows.

For a regular function F, we have

$$A[F(S)] = F'(S)A[S] + \frac{1}{2}F''(S)S^2\Sigma^2$$

and for a regular function of S and Σ

$$\begin{aligned} A[G(S,\Sigma)] = {} & G_1'(F,\Sigma)A[S] + G_2'(F,\Sigma)A[\Sigma] + \frac{1}{2}G_{11}''(F,\Sigma)\Gamma[S] \\ & + G_{12}''(F,\Sigma)\Gamma[S,\Sigma] + \frac{1}{2}G_{22}''(F,\Sigma)\Gamma[\Sigma]. \end{aligned}$$

By sharing $[0,T]$ into n subintervals and if the value of the option at time T is $F\big(S_T\big)$, even if we assume that the bias vanishes on the spot value, i.e. making $A[S] = 0$ in the above relation, we observe that the value of the option already at time $\frac{(n-1)T}{n}$ depends on both S and Σ.

If $F_t(x,y)$ is the function, which when taken at $\big(S_t, \Sigma_t\big)$ gives the value of the option, the transition equation between time $t+h$ and time t is:

$$\begin{aligned} F_t(S,\Sigma) = F_{t+h}(S,\Sigma) + \bigg\{ & \frac{\partial F_{t+h}}{\partial y}(S,\Sigma)A[\Sigma] + \frac{1}{2}\frac{\partial^2 F_{t+h}}{\partial x^2}(S,\Sigma)\Gamma[S] \\ & + \frac{\partial^2 F_{t+h}}{\partial x\partial y}(S,\Sigma)\Gamma[S,\Sigma] + \frac{1}{2}\frac{\partial^2 F_{t+h}}{\partial y^2}(S,\Sigma)\Gamma[\Sigma]\bigg\} h. \end{aligned}$$

In the above cases a), b), c) this yields a differential operator B in x and y such that

$$F_t = (I + hB)F_{t+h}$$

or

$$\frac{\partial F_t}{\partial t} + BF_t = 0.$$

It is apparent that, in addition to hypotheses on Γ such as a), b), c), a hypothesis is needed on $A[\Sigma]$. Then a pricing procedure can be conducted similar to that of Sections 2.1 and 2.2.

Concerning hedging, the presence of terms in $\frac{\partial}{dy}, \frac{\partial^2}{\partial x \partial y}, \frac{\partial^2}{\partial y^2}$ makes exact hedging impossible, in general, if the quantity Σ is not quoted in the market, we encounter the same questions as in the classical approach to stochastic volatility.

Comment. The argument pursued in Sections 2.1 and 2.2 is not classical; It cannot be reduced to the classical probabilistic reasoning which represents the asset price by means of a stochastic process and seeks, or rather assumes, an equivalent probability under which the stochastic process is a martingale.

Here instead, we start from hypotheses in terms of error structures that provide pricing for an infinitely-small time increment, and then global pricing. The question of hedging is split and depends on assumptions for the stochastic process of the asset price.

Let us sketch the complete mathematical framework of the approach. If the asset price is a process $S_t(\omega)$, we suppose that the instantaneous error on S_t is governed by an image error structure on $\mathbb{R}_+$ (in the scalar case) of the type

$$\Sigma_t = \big(\mathbb{R}_+, \mathcal{B}(\mathbb{R}_+), \mu_t, \mathbb{D}_t, \Gamma_t\big),$$

such that the associated generator A_t satisfies

$$A_t[I]\big(S_t(\omega)\big) = 0 \tag{28}$$

(if we were to omit the discounting procedure). Hence, the error structure Σ_t is moving and depends both on t and ω.

In adding the following hypothesis to (28)

$$\begin{cases} S_t \text{ is a continuous semi-martingale such that} \\ \dfrac{d\langle S, S\rangle_t}{dt} = \Gamma_t[I] \end{cases} \tag{29}$$

as shown in Sections 2.1 and 2.2, we once again find the classical case of a Markovian model with local volatility and exact hedging.

But hypotheses (28) and (29) may be split. Several possibilities are provided to the modeler or trader.

We may reasonably consider that the appropriate relation is rather an inequality

$$\frac{d\langle S, S\rangle_t}{dt} \leq \Gamma_t[I](S_t)$$

which represents, at best, the reality, Γ_t possibly involving errors other than the price temperature.

We may also consider that Γ_t is yielded by

$$\Gamma_t[I](x) = \sigma_t^2 x^2$$

where σ_t is the *implicit volatility* whereas $\frac{d\langle S,S\rangle_t}{S_t^2\,dt}$ represents the proportional quadratic variation density of the process S_t taken on the sample path, i.e. the *instantaneous historical volatility*. In this case pricing is performed by using market information on the quotation of options, while hedging is carried out using the process of the asset price itself, i.e. the only means available to the trader.

In what follows we will return to a more classical financial framework and use the tools of error calculus in order to study the sensitivity of models to hypotheses and of the results to the mistakes committed by the trader. We will of course start with the simplest model.

3 Error calculations on the Black–Scholes model

The ingredients of the Black–Scholes model are: a Brownian motion, two positive real parameters, the volatility and the interest rate, and the initial value of the asset. Starting with these quantities, the model computes

– the price of options

– the hedging, i.e., the composition of a portfolio simulating the option.

Two distinct issues arise concerning sensitivity.

1°) Studying the sensitivity of model outputs (option prices, hedges) to variations in the ingredients. For this topic the advantages of error calculus based upon Dirichlet forms are: to allow for a Lipschitzian calculus, to easily handle the propagation of errors through stochastic differential equations, and to consider errors on the Brownian motion itself.

2°) Studying the error on the result of a faulty hedging policy, when the trader misreads the right hypotheses in building his portfolio. This topic is different and will be tackled in due course.

3.1 Sensitivity of the theoretical pricing and hedging to errors on parameters and the asset price.

Notation. The interest rate of the bond is constant. The asset $(S_t)_{t\geq 0}$ is modeled as the solution to the equation $dS_t = S_t(\sigma\, dB_t + \mu\, dt)$. Theory actually reveals that the pricing and hedging formulae do not involve the drift coefficient μ. We therefore set $\mu = r$, i.e. working under the probability $\mathbb{P}$ such that $\tilde{S}_t = e^{-rt} S_t$, the discount stock price, is a martingale. For a European option with payoff $f(S_T)$, T fixed deterministic time, the value at time $t \in [0, T]$ of the option is $V_t = F\big(t, S_t, \sigma, r\big)$ with

$$F(t, x, \sigma, r) = e^{-r(T-t)} \int_{\mathbb{R}} f\left(x e^{\left(r - \frac{\sigma^2}{2}\right)(T-t) + \sigma y \sqrt{T-t}} \right) \frac{e^{-\frac{y^2}{2}}}{\sqrt{2\pi}}\, dy. \tag{30}$$

If f is Borel with linear growth, the function F is $\mathcal{C}^1$ in $t \in [0, T[$, $\mathcal{C}^2$ and Lipschitz in $x \in]0, \infty[$. We then set

$$\begin{aligned} \text{delta}_t &= \frac{\partial F}{\partial x}(t, S_t, \sigma, r) \\ \text{gamma}_t &= \frac{\partial^2 F}{\partial x^2}(t, S_t, \sigma, r) \\ \text{rho}_t &= \frac{\partial F}{\partial r}(t, S_t, \sigma, r). \end{aligned} \tag{31}$$

F satisfies the equations

$$\begin{cases} \dfrac{\partial F}{\partial t} + \dfrac{\sigma^2 x^2}{2} \dfrac{\partial^2 F}{\partial x^2} + r x \dfrac{\partial F}{\partial x} - r F = 0 \\ F(T, x, \sigma, r) = f(x). \end{cases} \tag{32}$$

Hypotheses. a) The error on $(B_t)_{t \geq 0}$ is represented by an Ornstein–Uhlenbeck possibly scaled error structure,

b) The errors on both the initial value S_0 and volatility σ are of the types discussed in Section 1. It may seem surprising to introduce an error on S_0, since it is usually assumed to be exactly known. As explained above, this approach entails a lack of accuracy on the time as to when the portfolio begins.

c) A constant proportional error is considered on the interest rate.

d) *A priori* laws are chosen on S_0, σ and r, but have not yet been specified (log-normal, exponential, uniform on an interval, etc.).

e) The random or randomized quantities $(B_t)_{t \geq 0}$, S_0, σ, r are assumed to be independent with uncorrelated errors. In other words, the quadratic error on a regular function $G\big((B_t)_{t \geq 0}, S_0, \sigma, r\big)$ will be

$$\begin{aligned} \Gamma\big[G(B, S_0, \sigma, r)\big] = \Gamma_{\mathrm{OU}}\big[G(\cdot, S_0, \sigma, r)\big](B) &+ G'^2_{S_0}(B, S_0, \sigma, r)\Gamma[S_0] \\ &+ G'^2_{\sigma}(B, S_0, \sigma, r)\Gamma[\sigma] + G'^2_{r}(B, S_0, \sigma, r)\Gamma[r]. \end{aligned}$$

Since $S_t = S_0 \exp\big\{\sigma B_t - \frac{\sigma^2}{2} t + rt\big\}$, we obtain by functional calculus on Γ

$$\frac{\Gamma[S_t]}{S_t^2} = \sigma^2 t \Gamma_{\mathrm{OU}}[B_1] + \frac{\Gamma[S_0]}{S_0^2} + (B_t - \sigma t)^2 \Gamma[\sigma] + t^2 \Gamma[r].$$

Here, Γ_{OU} is for the Ornstein–Uhlenbeck quadratic operator, and $\Gamma_{\mathrm{OU}}[B_1]$ is a scalar coefficient representing the size of the error on B.

We now consider a European option of payoff $f(S_T)$, where f is Lipschitz. By the independence hypotheses, the errors on B, S_0, σ, r can be handled separately. We denote $\Gamma_B, \Gamma_0, \Gamma_\sigma, \Gamma_r$ the corresponding quadratic operators.

3.1.1 Errors due to Brownian motion. Since B is present only in S_t, we have

$$\Gamma_B[V_t] = \left(\frac{\partial F}{\partial x}(t, S_t, \sigma, r)\right)^2 \Gamma_B[S_t],$$

thus

$$(33)\qquad \begin{cases} \Gamma_B[V_t] = \mathrm{delta}_t^2\, \Gamma_B[S_t] \\ \Gamma_B[V_s, V_t] = \mathrm{delta}_s\, \mathrm{delta}_t\, \Gamma_B[S_s, S_t] \end{cases}$$

with $\Gamma_B[S_s, S_t] = S_s S_t \sigma^2 (s \wedge t) \Gamma_{\mathrm{OU}}[B_1]$.

The following proposition shows that the error on B does not prevent the hedging portfolio from converging to the payoff if we assume the payoff function is Lipschitz.

Proposition VII.3. *If f is Lipschitz, V_t is in $\mathbb{D}_B$ and when $t \uparrow T$*

$$V_t = F(t, S_t, \sigma, r) \to f(S_T) \text{ in } \mathbb{D}_B \text{ and } \mathbb{P}\text{-a.s.}$$

$$\Gamma_B[V_t] = (\mathrm{delta}_t)^2 \Gamma_B[S_t] \to f'^2(S_T)\Gamma_B[S_T] \text{ in } L^1 \text{ and } \mathbb{P}\text{-a.s.}$$

Proof. First suppose $f \in \mathcal{C}^1 \cap \mathrm{Lip}$. By the relation

$$V_t = \mathbb{E}\big[e^{-r(T-t)} f(S_T) \mid \mathcal{F}_t\big]$$

it then follows that $V_t \to f(S_T)$ in L^p, $1 \le p < \infty$, and a.s.

A computation, to be performed in Section 4 within the more general framework of diffusion processes and which will not be repeated here, yields (see Chapter V, Section 2 for the definition of the sharp # and the hat ^)

$$V_t^\# = e^{-r(T-t)} \mathbb{E}\big[f'(S_T)S_T \mid \mathcal{F}_t\big]\sigma \hat{B}_t.$$

Thus

$$V_t^\# \to f'(S_T)S_T\sigma\hat{B}_T \quad \text{in } L^2\big(\mathbb{P}, L^2(\hat{\Omega}, \hat{\mathbb{P}})\big),$$

and thanks to $f(S_T)^\# = f'(S_T)S_T\sigma\hat{B}_T$ we obtain

$$V_t \to f(S_T) \quad \text{in } \mathbb{D}_B \text{ and } \mathbb{P}\text{-a.s.}$$

and

$$\Gamma_B[V_t] = e^{-2r(T-t)}\big(\mathbb{E}\big[f'(S_T)S_T \mid \mathcal{F}_t\big]\big)^2\sigma^2 t \to f'^2(S_T)\Gamma_B[S_T]$$

in L^1 and $\mathbb{P}$-a.s.

The case of only Lipschitz hypotheses stems from a special property of one-variable functional calculus in error structures (see Chapter III, Section 3). The preceding argument remains valid. ◇

Let us now investigate the error due to B on the hedging portfolio. We assume f and f' in $\mathcal{C}^1 \cap \mathrm{Lip}$. $\tilde{S}_t = e^{-rt} S_t$ is the discount asset price. The hedging equation then is

$$e^{-rt} F(t, S_t, \sigma, r) = F(0, S_0, \sigma, r) + \int_0^t H_s \, d\tilde{S}_s,$$

where the adapted process H_t is the quantity of stock in the portfolio:

$$H_t = \text{delta}_t = \frac{\partial F}{\partial x}(t, S_t, \sigma, r) = e^{-r(T-t)} \mathbb{E}\big[f'(S_T) S_T \mid \mathcal{F}_t\big] \frac{1}{S_t}.$$

By the same method as for V_t we obtain

$$\text{(34)} \qquad \begin{cases} \Gamma_B[H_t] = (\text{gamma}_t)^2 \, \Gamma_B[S_t] \\ \Gamma_B[H_s, H_t] = \text{gamma}_s \, \text{gamma}_t \Gamma_B[S_s, S_t]. \end{cases}$$

Proposition VII.4. *If $f, f' \in \mathcal{C}^1 \cap \mathrm{Lip}$, then $H_t \in \mathbb{D}$ and as $t \uparrow T$*

$$\begin{aligned} H_t &\to f'(S_T) && \text{in } \mathbb{D}_B \text{ and a.s.} \\ \Gamma_B[H_t] &\to f''^2(S_T)\Gamma_B[S_T] && \text{in } L^1(\mathbb{P}) \text{ and a.s.} \end{aligned}$$

3.1.2 Error due to volatility. Γ_σ denotes the quadratic error operator on σ; let us denote D_σ the corresponding gradient with values in $\mathcal{H} = \mathbb{R}$. We suppose the payoff function f in $\mathcal{C}^1 \cap \mathrm{Lip}$. Since $V_t = F(t, S_t, \sigma, r)$, we have

$$D_\sigma[V_t] = \left(F'_x(t, S_t, \sigma, r) \frac{\partial S_t}{\partial \sigma} + F'_\sigma(t, S_t, \sigma, r) \right) D_\sigma[I].$$

Remarking that (30) yields

$$\frac{\partial F}{\partial \sigma} = (T-t)\sigma x^2 \frac{\partial^2 F}{\partial x^2},$$

we obtain

$$\text{(35)} \qquad \begin{cases} D_\sigma[V_t] = \big(S_t(B_t - \sigma t)\text{delta}_t + (T-t)\sigma S_t^2 \, \text{gamma}_t\big) D_\sigma[I] \\ \Gamma_\sigma[V_t] = \big(S_t(B_t - \sigma t)\text{delta}_t + (T-t)\sigma S_t^2 \, \text{gamma}_t\big)^2 \Gamma_\sigma[I]. \end{cases}$$

In order to study the bias due to an error on σ, suppose that the error structure on σ is such that the identity map I belongs to $(\mathcal{D}A_\sigma)_{\text{loc}}$ (in a suitable sense which we have not necessarily defined but which is clear in most applications). We then have

$$A_\sigma[V_t] = \frac{dV_t}{d\sigma} A_\sigma[I] + \frac{1}{2} \frac{d^2 V_t}{d\sigma^2} \Gamma_\sigma[I],$$

which, for the pricing at $t = 0$, yields

$$(36)\quad \begin{cases} \text{(i)} & D_\sigma[V_0] = T\sigma S_0^2 \,\text{gamma}_0\, D_\sigma[I] \\ \text{(ii)} & \Gamma_\sigma[V_0] = T^2\sigma^2 S_0^4 \,\text{gamma}_0^2\, \Gamma_\sigma[I] \\ \text{(iii)} & A_\sigma[V_0] = T\sigma S_0^2 \,\text{gamma}_0\, A_\sigma[I] + \frac{1}{2}\frac{d^2V_0}{d\sigma^2}\Gamma_\sigma[I]. \end{cases}$$

Equations (36)(i) and (36)(ii) express well-known facts: for example they show that for European options of payoffs $f_{(1)}(S_T)$ and $f_{(2)}(S_T)$, an option with payoff $a_1 f_{(1)}(S_T)+ a_2 f_{(2)}(S_T)$ would have a value at time 0 insensitive to σ (i.e. $\Gamma_\sigma[V_0] = 0$) once $a_1\,\text{gamma}_0^{(1)} + a_2\,\text{gamma}_0^{(2)} = 0$.

Equation (36)(ii) also shows that if the payoff function is convex [resp. concave] the price of the option increases [resp. decreases] as σ increases. (This property has been demonstrated to remain true in diffusion models when volatility depends solely on stockprice level, see El Karoui et al. [1998].)

The relation (36)(iii) highlights *nonlinear* phenomena. The bias on pricing V_0 is not necessarily positive, even when the payoff function is convex. Let us consider the case where the error structure for σ is such that $I \in (\mathcal{D}A_\sigma)_{\text{loc}}$ and $A_\sigma[I](\sigma_0) = 0$. In other words, the error on σ is centered when σ displays the value σ_0. We then have

$$(37)\qquad A_\sigma[V_0](\sigma_0) = \frac{1}{2}\frac{d^2V_0}{d\sigma^2}\Gamma_\sigma[I](\sigma_0).$$

The interpretation is as follows: although the function $\sigma \to V_0(\sigma)$ is increasing for a convex payoff function, this function is not linear and a centred error on σ may yield a negative bias on V_0. In such a situation, when $\frac{d^2V_0}{d\sigma^2}(\sigma_0) < 0$, if the trader considers his error on σ to be statistically centered, his pricing will, on average, be undervalued.

Regarding convergence of the hedging portfolio, from (35), we have:

$$-f(S_T)] = \big\{\mathbb{E}\big[e^{-r(T-t)} f'(S_T)(B_T - \sigma T)S_T \mid \mathcal{F}_t\big] - f'(S_T)(B_T - \sigma T)S_T\big\}^2 \Gamma_\sigma[I],$$

expectation only concerns Brownian motion B. This indicates the same result of convergence for the error due to σ as for the errors due to B: errors do not prevent the portfolio from converging to the payoff with errors tending to zero as $t \to T$.

Concerning the error due to r, similarly we have

$$\Gamma_r[V_t] = \left\{F'_x(t, S_t, \sigma, r)\frac{\partial S_t}{\partial r} + F'_r(t, S_t, \sigma, r)\right\}^2 \Gamma_r[I]$$

and

$$\Gamma_r[V_t] = \big\{tS_t\,\text{delta}_t + \text{rho}_t\big\}^2 \Gamma_r[I].$$

As a consequence, given several options of payoffs $f_{(i)}(S_T)$, $i = 1, \dots, k$, the option of payoff $\sum_i a_i f_{(i)}(S_T)$ has a value at time 0 insensitive to both σ and r (i.e., $\Gamma_\sigma[V_0] = \Gamma_r[V_0] = 0$) if the vector $a = (a_i)$ is orthogonal to the two vectors $(\text{gamma}_0^{(i)})$ and $(\text{rho}_0^{(i)})$.

The preceding computations easily show that in the Black–Scholes model, if U_1 and U_2 are two random variables taken from among the following quantities defined at a fixed instant t: S_t, $V_t(f_1)$, $V_t(f_2)$, $H_t(f_1)$, $H_t(f_2)$, then the matrix $\Gamma\big[U_i, U_j\big]$ is singular: the errors on these quantities are linked. This finding stems from the fact that the law of, for example, the pair $\big(V_t(f_1), V_2(f_2)\big)$ is carried by the λ-parameterized curve

$$y = \exp -r(T-t)P_{T-t}f_1(\lambda)$$
$$x = \exp -r(T-t)P_{T-t}f_2(\lambda),$$

where (P_t) is the transition semigroup of (S_t). The same phenomenon occurs in any other general Markovian model.

On the contrary, the random quantities involving several different instants generally have non-linked errors. Thus for example, if $U_1 = S_T$ and $U_2 = \int_0^T e^{-s} H_s S_s\, ds$ (discounted immobilization of the portfolio) the matrix $\Gamma\big[U_i, U_j\big]$ is a.s. regular as long as f is not constant (see Chapter III, Section 3), hence, by the absolute continuity criterion, the law of the pair $\big(S_T, \int_0^T e^{-s} H_s S_s\, ds\big)$ possesses a density.

3.2 Errors uniquely due to the trader. In the preceding discussion we considered that the scalar parameters σ, S_0, r and Brownian motion B have been erroneous, i.e. containing intrinsic inaccuracies, and we studied the corresponding errors on derived model quantities.

In particular, the error on σ concerned all uses of σ: the turbulence of S_t together with the hedging formulae. Hedging followed the perturbations of σ, which implied convergence of the hedging portfolio to the payoff provided the functions were smooth enough.

We now suppose that only the trader mishandles the pricing and hedging when reading σ to calculate V_0 and obtain H_t. Nonetheless we assume that stock price S_t is not erroneous neither at $t = 0$ nor thereafter.

3.2.1 Pricing error. If the trader chooses an incorrect σ and uses the Black–Scholes model to price a European option of payoff $f(S_T)$ at time T, he is committing a pricing error, i.e. on the initial value of his hedging portfolio.

The calculation is a simple derivation of the Black–Scholes formula with respect to σ. We have already derived the following:

$$D_\sigma[V_0] = F'_\sigma\big(0, S_0, \sigma, r\big) D_\sigma[I];$$

(i) $D_\sigma[V_0] = T\sigma S_0^2 \,\mathrm{gamma}_0\, D_\sigma[I]$,

(ii) $\Gamma_\sigma[V_0] = T^2\sigma^2 S_0^4 \,\mathrm{gamma}_0^2\, \Gamma_\sigma[I]$,

(iii) $A_\sigma[V_0] = T\sigma S_0^2 \operatorname{gamma}_0 A_\sigma[I] + \dfrac{1}{2}\dfrac{d^2V_0}{d\sigma^2}\Gamma_\sigma[I].$

This calculation is especially interesting for nonstandard options not quoted on the markets and sold over the counter. For quoted options, the trader has no real choice for the pricing. On the other hand, for the hedging, he must choose σ in order to apply the Black–Scholes model formulae.

3.2.2 Hedging error. We assume that the market follows a Black–Scholes model and that for hedging a European option of payoff $f(S_T)$, the trader has built a portfolio whose initial value is correct (and given by the market) which consists however of an incorrect quantity of assets.

The notation is as follows:

$$\begin{aligned}\tilde{F}(t,x,\sigma,r) &= e^{-rt}F\big(t,xe^{rt},\sigma,r\big)\\ \tilde{S}_t &= e^{-rt}S_t.\end{aligned}$$

The hedging equation is

$$\tilde{V}_t = V_0 + \int_0^t \frac{\partial \tilde{F}}{\partial x}\big(s,\tilde{S}_s,\sigma,r\big)\,d\tilde{S}_s.$$

We suppose that the portfolio constituted by the trader has the discounted value

$$\tilde{P}_t = V_0 + \int_0^t \frac{\partial \tilde{F}}{\partial x}\big(s,\tilde{S}_s,\sigma,r\big)d\tilde{S}_s,$$

where only the argument σ of $\frac{\partial \tilde{F}}{\partial x}(s,\tilde{S}_s,\sigma,r)$ is erroneous. The functional calculus then yields

$$D_\sigma\big[\tilde{P}_t\big] = \int_0^t \frac{\partial^2 \tilde{F}}{\partial\sigma\,\partial x}\big(s,\tilde{S}_s,\sigma,r\big)\,d\tilde{S}_s \cdot D_\sigma[I].$$

To evaluate this stochastic integral, we use Itô's formula in supposing F to be sufficiently regular.

$$\begin{aligned}\frac{\partial \tilde{F}}{\partial \sigma}\big(t,\tilde{S}_t,\sigma,r\big) = \frac{\partial \tilde{F}}{\partial \sigma}(0,\tilde{S}_0,\sigma,r) &+ \int_0^t \frac{\partial^2 \tilde{F}}{\partial\sigma\,\partial x}\big(s,\tilde{S}_s,\sigma,r\big)\,d\tilde{S}_s\\ &+ \int_0^t \frac{\partial^2 \tilde{F}}{\partial\sigma\,\partial t}\big(s,\tilde{S}_s,\sigma,r\big)\,ds + \frac{1}{2}\int_0^t \frac{\partial^3 \tilde{F}}{\partial\sigma\,\partial x^2}(s,\tilde{S}_s,\sigma,r)\sigma^2\tilde{S}_s^2\,ds.\end{aligned}$$

The function $\tilde{F}$ satisfies

$$\begin{cases}\dfrac{\partial \tilde{F}}{\partial t} + \dfrac{1}{2}x^2\sigma^2\dfrac{\partial^2 \tilde{F}}{\partial x^2} = 0\\[2ex] \tilde{F}\big(T,xe^{-rT},\sigma,r\big) = e^{-rT}f(x).\end{cases}$$

Hence $\frac{\partial \tilde{F}}{\partial \sigma}$ satisfies

$$\begin{cases} \dfrac{\partial^2 \tilde{F}}{\partial \sigma \partial t} + \dfrac{1}{2} x^2 \sigma^2 \dfrac{\partial^3 \tilde{F}}{\partial \sigma \partial x^2} + x^2 \sigma \dfrac{\partial^2 \tilde{F}}{\partial x^2} = 0 \\ \dfrac{\partial \tilde{F}}{\partial \sigma}\big(T, xe^{-rT}, \sigma, r\big) = 0. \end{cases}$$

From this development we can draw the following:

$$\begin{aligned} D_\sigma[\tilde{P}_t] &= \left\{ \frac{\partial \tilde{F}}{\partial \sigma}(t, \tilde{S}_t, \sigma, r) - \frac{\partial \tilde{F}}{\partial \sigma}(0, \tilde{S}_0, \sigma, r) + \int_0^t \tilde{S}_s^2 \sigma \frac{\partial^2 \tilde{F}}{\partial x^2}(s, \tilde{S}_s, \sigma, r) ds \right\} D_\sigma[I] \\ &= \left\{ e^{-rt} \frac{\partial F}{\partial \sigma}(t, S_t, \sigma, r) - \frac{\partial F}{\partial \sigma}(0, S_0, \sigma, r) \right. \\ &\quad \left. + \int_0^t e^{-rs} S_s^2 \sigma \frac{\partial^2 F}{\partial x^2}(s, S_s, \sigma, r)\, ds \right\} D_\sigma[I] \end{aligned}$$

and for $t = T$

$$D_\sigma[\tilde{P}_T] = \left\{ -\frac{\partial F}{\partial \sigma}(0, S_0, \sigma, r) + \int_0^T e^{-rs} S_s^2 \sigma \,\mathrm{gamma}_s\, ds \right\} D_\sigma[I], \tag{38}$$

or, setting $P_T = e^{rT} \tilde{P}_T$,

$$D_\sigma[P_T] = e^{rT} D_\sigma[\tilde{P}_T] \tag{39}$$

$$\Gamma_\sigma[P_T] = e^{2rT} \left\{ \int_0^T e^{-rs} S_s^2 \sigma \,\mathrm{gamma}_s\, ds - \frac{\partial F}{\partial \sigma}(0, S_0, \sigma, r) \right\}^2 \Gamma_\sigma[I]. \tag{40}$$

According to (38)–(40), the (algebraic) benefit due to the hedging error is

$$\Delta_\sigma = e^{rT} \left\{ \int_{[0,T]} e^{-rs} S_s^2 \sigma \,\mathrm{gamma}_s \big(ds - T\delta_0(s)\big) \right\} D_\sigma[I]$$

where δ_0 is the Dirac measure at 0.

We note that if the path of the process

$$e^{-rs} S_s^2(\omega) \frac{\partial^2 F}{\partial x^2}\big(s, S_s(\omega), \sigma, r\big)$$

varies marginally, such that its evolution on $[0, T]$ satisfies

$$\int_0^T e^{-rs} S_s^2(\omega) \frac{\partial^2 F}{\partial x^2}\big(s, S_s(\omega), \sigma, r\big)\, ds \,\#\, T S_0^2(\omega) \frac{\partial^2 F}{\partial x^2}\big(0, S_0, \sigma, r\big),$$

then the error on the hedging almost vanishes. This result expresses a stability property of the Black–Scholes hedging.

In returning to partial derivatives with respect to time using $\frac{\partial \tilde{F}}{\partial t} + \frac{1}{2}\sigma^2 x^2 \frac{\partial^2 \tilde{F}}{\partial x^2} = 0$ we obtain

$$D_\sigma[\tilde{P}_T] = \frac{2}{\sigma}\int_0^T \left\{\frac{\partial \tilde{F}}{\partial t}(0, S_0, \sigma, r) - \frac{\partial \tilde{F}}{\partial t}(s, \tilde{S}_s, \sigma, r)\right\} ds \cdot D_\sigma[I].$$

We observe that the sign of the process

$$\frac{\partial \tilde{F}}{\partial t}(0, S_0, \sigma, r) - \frac{\partial \tilde{F}}{\partial t}(s, \tilde{S}_s, \sigma, r)$$

determines whether a positive error on σ results in a benefit or deficit.

3.2.3 Error on the interest rate. If the trader commits an error on the rate r independently of that committed on σ, we can treat this error separately.

a) For the pricing we have

$$\begin{aligned} D_r[V_0] &= F_r'(0, S_0, \sigma, r) D_r[I] \\ &= \big(S_0 T \,\text{delta}_0 - T V_0\big) D_r[I] \end{aligned}$$

$$\Gamma_r[V_0] = \big(S_0 T \,\text{delta}_0 - T V_0\big)^2 \Gamma_r[I]$$

$$A_r[V_0] = \big(S_0 T \,\text{delta}_0 - T V_0\big) A_r[I] + \frac{1}{2}\frac{\partial^2 V_0}{\partial r^2}\Gamma_r[I].$$

b) For the hedging,

$$\begin{aligned} D_r[\tilde{P}_t] &= \int_0^t \frac{\partial^2 \tilde{F}}{\partial r \partial x}(s, \tilde{S}_s, \sigma, r) d\tilde{S}_s \cdot D_r[I] \\ &= \left\{\frac{\partial \tilde{F}}{\partial r}\big(t, \tilde{S}_t, \sigma, r\big) - \frac{\partial \tilde{F}}{\partial r}(0, S_0, \sigma, r)\right\} D_r[I]. \end{aligned}$$

Thus for $t = T$, using the equation satisfied by F,

$$D_r[\tilde{P}_T] = \left(T e^{-rT}\big(S_T f'(S_T) - f(S_T)\big) - \frac{\partial F}{\partial r}(0, S_0, \sigma, r)\right) D_r[I],$$

or, equivalently,

$$D_r[\tilde{P}_T] = T\big\{e^{-rT}\big(S_T \,\text{delta}_T - f(S_T)\big) - \big(S_0 \,\text{delta}_0 - V_0\big)\big\} D_r[I].$$

These formulae show that the (algebraic) benefit due to an incorrect value of the interest rate of the only hedging is

$$\Delta_r = T e^{rT}\big\{e^{-rT}\big(S_T \,\text{delta}_T - f(S_T)\big) - \big(S_0 \,\text{delta}_0 - V_0\big)\big\} D_r[I].$$

(We have used $P_T = e^{rT}\tilde{P}_T$ without considering r to be erroneous since the error on r concerns only the composition of the hedging portfolio.) We can see that the increment of the process $e^{-rt}\big(S_t\,\text{delta}_t - f(S_t)\big)$ between 0 and T is what determines whether a positive error on r produces an advantage or a loss. Unlike the case of σ, it is not the whole path of the process that matters, but only the initial and final values.

Exercise (Error for floating Black–Scholes). Let us evaluate, for purpose of clarity, the gap in hedging when the trader bases his portfolio at each time on the implicit volatility. This calculation does not involve any error structure.

We suppose that for the hedge of a European option quoted on the markets, a trader uses the market price and the implicit volatility given by the Black–Scholes formula and then constructs at each time its portfolio as if this volatility were constant. Does this procedure reach the payoff at the exercise time?

Suppose the interest rate to be zero for the sake of simplicity.

Let V_t be the price of the option. The implicit volatility σ_t^i is then deduced by

$$V_t = F\big(t, S_t, \sigma_t^i, 0\big).$$

If σ_t^i is a process with finite variation, what is the simplest hypothesis after a constant, and if S_t is a continuous semimartingale, using $\frac{\partial F}{\partial t} + \frac{1}{2}\sigma^2 x^2 \frac{\partial^2 F}{\partial x^2} = 0$ we have

$$\begin{aligned} dV_s &= \frac{\partial F}{\partial x}\big(s, S_s, \sigma_s^i, 0\big)\, dS_s + \frac{1}{2}\frac{\partial^2 F}{\partial x^2}\big(s, S_s, \sigma_s^i, 0\big)\big(d\langle S, S\rangle_s - \sigma_s^i S_s^2\, ds\big) \\ &\quad + \frac{\partial F}{\partial \sigma}\big(s, S_s, \sigma_s^i, 0\big)\, d\sigma_s^i. \end{aligned}$$

Let us suppose the process σ_t^i to be absolutely continuous and let us introduce the historical volatility σ_t^h, defined by

$$d\langle S, S\rangle_t = S_t^2 \big(\sigma_t^h\big)^2\, dt.$$

Using $\frac{\partial F}{\partial \sigma} = (T-t)\sigma x^2 \frac{\partial^2 F}{\partial x^2}$ we obtain

$$\begin{aligned} V_T &= V_0 + \int_0^T \frac{\partial F}{\partial x}\big(s, S_s, \sigma_s^i, 0\big)\, dS_s \\ &\quad + \frac{1}{2}\int_0^T \frac{\partial^2 F}{\partial x^2}\big(s, S_s, \sigma_s^i, 0\big)\left((\sigma_s^h)^2 - (\sigma_s^i)^2 + 2(T-s)\sigma_s^i \frac{d\sigma_s^i}{ds}\right) S_s^2 ds. \end{aligned}$$

We can observe that for a convex payoff, the result of this procedure compared with the market price V_T $(= f(S_T))$ may be controlled at each time by the sign of the expression

$$(\sigma_s^h)^2 - (\sigma_s^i)^2 + (T-s)\frac{d(\sigma_s^i)^2}{ds} = (\sigma_s^h)^2 + \frac{d}{ds}\big((T-s)(\sigma_s^i)^2\big).$$

4 Error calculations for a diffusion model

We will now extend the preceding study to the case of a Markovian model in which the asset X_t is governed by a stochastic differential equation of the type:

$$dX_t = X_t\sigma\big(t, X_t\big)\,dB_t + X_t r(t)\,dt.$$

We first study the sensitivity of the theoretical pricing and the hedging to changes in model data. To begin with, we suppose the Brownian motion to be erroneous and then consider the case in which the function $(t,x) \to \sigma(t,x)$ is erroneous. This setup leads to studying the sensitivity of the solution of a stochastic differential equation to an error on its functional coefficients.

Later on we will study the consequences on the pricing and hedging of an error due uniquely to the trader.

4.1 Sensitivity of the theoretical model to errors on Brownian motion. The stock price is assumed to be the solution to the equation

$$dX_t = X_t\sigma\big(t, X_t\big)\,dB_t + X_t r(t)\,dt. \tag{41}$$

The interest rate is deterministic and the function $\sigma(t,x)$ will be supposed bounded with a bounded derivative in x uniformly in $t \in [0,T]$. The probability is a martingale measure, such that if $f(X_T)$ is the payoff of a European option, its value at time t is

$$V_t = \mathbb{E}\left[\exp\left(-\int_t^T r(s)\,ds\right) f(X_T) \mid \mathcal{F}_t\right] \tag{42}$$

where $(\mathcal{F}_t)$ is the Brownian filtration. The hedging portfolio is given by the adapted process H_t, which satisfies

$$\tilde{V}_t = V_0 + \int_0^t H_s\,d\tilde{X}_s, \tag{43}$$

where $\tilde{V}_t = \exp\big(-\int_0^t r(s)\,ds\big)V_t$ and $\tilde{X}_t = \exp\big(-\int_0^t r(s)\,ds\big)X_t$.

Hypotheses for the errors on B. We suppose here that the Brownian motion $\big(B_t\big)_{t\geq 0}$ is erroneous (see Chapter VI, Section 2). The Ornstein–Uhlenbeck structure on the Wiener space is invariant by translation of the time:

$$\Gamma\left[\int_0^\infty u(s)\,dB_s\right] = \Gamma\left[\int_0^\infty u(s+h)\,dB_s\right] \quad \forall u \in L^2(\mathbb{R}_+).$$

In order to allow for a more general study, we will suppose that the Wiener space is equipped with a weighted Ornstein–Uhlenbeck structure (W.O.U.-structure).

Let α be a function on $\mathbb{R}_+$ such that $\alpha(x) \geq 0 \ \forall x \in \mathbb{R}_+$ and $\alpha \in L^1_{\text{loc}}(\mathbb{R}_+, dx)$. The W.O.U.-structure associated with α is then defined as the generalized Mehler-type structure associated with the semigroup on $L^2(\mathbb{R}_+)$:

$$p_t u = e^{-\alpha t} u.$$

This error structure satisfies

$$\Gamma\left[\int_0^\infty u(s)\, dB_s\right] = \int_0^\infty \alpha(s) u^2(s)\, ds$$

for $u \in \mathcal{C}_K(\mathbb{R}_+)$. It is the mathematical expression of a perturbation of the Brownian path

$$\omega(s) = \int_0^s dB_u \to \int_0^s e^{-\frac{\alpha(u)}{2}\varepsilon}\, dB_u + \int_0^s \sqrt{1 - e^{-\alpha(u)\varepsilon}}\, d\hat{B}_u,$$

where $\hat{B}$ is an independent Brownian motion.

This structure possesses the following gradient:

$$D\colon \mathbb{D} \to L^2(\mathbb{P}, \mathcal{H}) \quad \text{where } \mathcal{H} = L^2(\mathbb{R}_+, dt);$$

- $D\left[\int u(s)\, dB_s\right](t) = \sqrt{\alpha(t)}\, u(t) \quad \forall u \in L^2(\mathbb{R}_+, (1+\alpha)\, dt)$,
- if H_t is a regular adapted process

$$D\left[\int H_s\, dB_s\right](t) = \sqrt{\alpha(t)} H_t + \int D[H_s](t)\, dB_s.$$

We will also use the sharp, which is a particular gradient with $\mathcal{H} = L^2(\hat{\Omega}, \hat{\mathcal{A}}, \hat{\mathbb{P}})$ defined by

$$\left(\int_0^\infty u(s)\, dB_s\right)^\# = \int_0^\infty \sqrt{\alpha(s)}\, u(s)\, d\hat{B}_s, \quad u \in L^2(\mathbb{R}_+, (1+\alpha)dt),$$

which satisfies the chain rule and for a regular adapted process H

$$\left(\int_0^\infty H_s\, dB_s\right)^\# = \int_0^\infty \sqrt{\alpha(s)} H_s\, d\hat{B}_s + \int_0^\infty H_s^\#\, dB_s.$$

We require the two following lemmas, whose demonstrations are relatively straightforward.

Lemma VII.5. *The conditional expectation operator* $\mathbb{E}[\cdot \mid \mathcal{F}_t]$ *maps* $\mathbb{D}$ *into* $\mathbb{D}$*; it is an orthogonal projector in* $\mathbb{D}$ *and its range is an error sub-structure* (*close sub-vector space of* $\mathbb{D}$ *preserved by Lipschitz functions*).

Lemma VII.6. *Let Γ_t be defined from Γ by*

$$\Gamma_t\left[\int u(s)\,dB_s\right]=\Gamma\left[\int 1_{[0,t]}(s)u(s)\,dB_s\right]$$

and let $U \to U^{\#t}$ be the sharp operator associated with Γ_t, then for $U \in \mathbb{D}$

$$\left(\mathbb{E}[U \mid \mathcal{F}_t]\right)^{\#}=\mathbb{E}\left[U^{\#t} \mid \mathcal{F}_t\right].$$

We can now undertake the study of error propagation.

Propagation of an error on *B*. We proceed as follows. From the equation

$$X_t = X_0 + \int_0^t X_s\sigma(s, X_s)\,dB_s + \int_0^t X_s r(s)\,ds$$

we draw

$$\begin{aligned} X_t^{\#} &= \int_0^t \big(\sigma(s, X_s) + X_s\sigma_x'(s, X_s)\big)X_s^{\#}\,dB_s \\ &\quad + \int_0^t \sqrt{\alpha(s)}X_s\sigma(s, X_s)\,d\hat{B}_s + \int_0^t X_s^{\#}r(s)\,ds. \end{aligned} \tag{44}$$

This equation can be solved by means of a probabilistic version of the method of the constant variation:

If we set

$$\begin{cases} K_t = \sigma(t, X_t) + X_t\sigma_x'(t, X_t) \\ M_t = \exp\left\{\int_0^t K_s\,dB_s - \frac{1}{2}\int_0^t K_s^2\,ds + \int_0^t r(s)\,ds\right\}, \end{cases} \tag{45}$$

we have

$$X_t^{\#} = M_t \int_0^t \frac{\sqrt{\alpha(s)}X_s\sigma(s, X_s)}{M_s}\,d\hat{B}_s, \tag{46}$$

as is easily verified using Itô calculus. The effect of the error on $(B_t)_{t\geq 0}$ on the process $(X_t)_{t\geq 0}$ is given by

$$\Gamma[X_t] = M_t^2 \int_0^t \frac{\alpha(s)X_s^2\sigma^2(s, X_s)}{M_s^2}\,ds$$

$$\Gamma[X_s, X_t] = M_s M_t \int_0^{s\wedge t} \frac{\alpha(u)X_u^2\sigma^2(u, X_u)}{M_u^2}\,du.$$

Error on the value of the option. Let us suppose $f \in \mathcal{C}^1 \cap \text{Lip}$ (as usual in error structures, $\mathcal{C}^1 \cap \text{Lip}$ hypotheses are needed for functions of several arguments and Lipschitz hypotheses are sufficient when calculations concern a single argument). Let us define

$$Y = \exp\left(-\int_t^T r(s)\,ds\right) f(X_T).$$

In order to compute $\left(\mathbb{E}[Y \mid \mathcal{F}_t]\right)^{\#}$ we apply Lemma VII.6:

$$Y^{\#t} = \exp\left(-\int_t^T r(s)\,ds\right) f'(X_T) X_T^{\#t}$$

and

$$\begin{aligned}\left(\mathbb{E}[Y \mid \mathcal{F}_t]\right)^{\#} &= \exp\left(-\int_t^T r(s)\,ds\right) \mathbb{E}\big[f'(X_T) X_T^{\#t} \mid \mathcal{F}_t\big] \\ &= \exp\left(-\int_t^T r(s)\,ds\right) \mathbb{E}\big[f'(X_T) M_T \mid \mathcal{F}_t\big] \int_0^t \frac{\sqrt{\alpha(s)} X_s \sigma(s, X_s)}{M_s}\, d\hat{B}_s.\end{aligned}$$

Lemma VII.6 yields

$$\begin{aligned}\Gamma[V_t] &= \Gamma\big[\mathbb{E}[Y \mid \mathcal{F}_t]\big] \\ &= \exp\left(-2\int_t^T r(s)\,ds\right) \left(\mathbb{E}\big[f'(X_T) M_T \mid \mathcal{F}_t\big]\right)^2 \int_0^t \frac{\alpha(s) X_s^2 \sigma^2(s, X_s)}{M_s^2}\, ds.\end{aligned} \tag{47}$$

This also yields the cross-error of V_t and V_s, which is useful for computing errors on random variables such as $\int_0^T h(s)\,dV_s$ or $\int_0^T V_s h(s)\,ds$

(48)

$$\begin{aligned}\Gamma\big[V_s, V_t\big] = \exp&\left(-\int_s^T r(u)\,du - \int_t^T r(v)\,dv\right) \\ &\mathbb{E}\big[f'(X_T) M_T \mid \mathcal{F}_s\big] \mathbb{E}\big[f'(X_T) M_T \mid \mathcal{F}_t\big] \int_0^{s \wedge t} \frac{\alpha(u) X_u^2 \sigma^2(u, X_u)}{M_u^2}\, du.\end{aligned}$$

With our hypotheses and as $t \uparrow T$,

$$\Gamma[V_t] \to f'^2(X_T) M_T^2 \int_0^T \frac{\alpha(s) X_s^2 \sigma^2(X_s)}{M_s^2}\, ds = f'^2(X_T)\Gamma[X_T] = \Gamma[f(X_T)]$$

in $L^1(\mathbb{P})$ and a.s.

Error on the hedging portfolio. In order now to treat H_t, let us first remark that H_t is easily obtained using the Clark formula. For this purpose let us return to the classical Ornstein–Uhlenbeck framework ($\alpha(t) \equiv 1$) until formula (49). Relations (42) and (43) give

$$H_t \exp\left(-\int_0^t r(s)\,ds\right) X_t\sigma(t, X_t) = D_{\mathrm{ad}}\left[\exp\left(-\int_0^T r(s)\,ds\right) f(X_T)\right],$$

where D_{ad} is the adapted gradient defined via

$$D_{\mathrm{ad}}[Z](t) = \mathbb{E}\big[DZ(t) \mid \mathcal{F}_t\big].$$

Since

$$D\left[\exp\left(-\int_0^T r(s)\,ds\right) f(X_T)\right] = \exp\left(-\int_0^T r(s)\,ds\right) f'(X_T)(DX_T)(t)$$

we have, from the computation performed for V_t

$$\begin{aligned}
&D_{\mathrm{ad}}\left[\exp\left(-\int_0^T r(s)\,ds\right) f(X_T)\right](t)\\
&\quad= \exp\left(-\int_0^T r(s)\,ds\right) \mathbb{E}\big[f'(X_T)M_T \mid \mathcal{F}_t\big] \frac{X_t\sigma(t, X_t)}{M_t}.
\end{aligned}$$

Thus

$$H_t = \exp\left(-\int_t^T r(s)\,ds\right) \mathbb{E}\big[f'(X_T)M_T \mid \mathcal{F}_t\big] \frac{1}{M_t}. \tag{49}$$

Now supposing f and $f' \in \mathcal{C}^1 \cap \mathrm{Lip}$, applying the same method as that used for obtaining $\Gamma[V_t]$ leads to yields

$$\begin{aligned}
\Gamma[H_t] = \exp\left(-2\int_t^T r(s)\,ds\right) &\left(\mathbb{E}\left[\frac{M_T}{M_t}\big(f''(X_T)M_T + f'(X_T)Z_t^T\big) \mid \mathcal{F}_t\right]\right)^2\\
&\int_0^t \frac{\alpha(u)X_u^2\sigma(u, X_u)}{M_u^2}\,du
\end{aligned} \tag{50}$$

with

$$\begin{aligned}
Z_t^T &= \int_t^T L_s\,dB_s - \int_t^T K_s L_s M_s\,ds\\
K_s &= \sigma(s, X_s) + X_s\sigma_x'(s, X_s)\\
L_s &= 2\sigma_x'(s, X_s) + X_s\sigma_{x^2}''(s, X_s).
\end{aligned}$$

We introduce the following notation which extends the Black–Scholes case:

$$\text{delta}_t = H_t = \exp\left(-\int_t^T r(s)\,ds\right)\mathbb{E}\big[f'(X_T)M_T \mid \mathcal{F}_t\big]\frac{1}{M_t}$$

$$\text{gamma}_t = \exp\left(-\int_t^T r(s)\,ds\right)\mathbb{E}\left[\frac{M_T^2}{M_t^2}f''(X_T) + \frac{M_T}{M_t^2}f'(X_T)Z_t^T \mid \mathcal{F}_t\right].$$

We can now summarize the formulae for this diffusion case as follows:

$$\begin{aligned}
V_t^\# &= \text{delta}_t\, X_t^\# \\
H_t^\# &= \text{gamma}_t\, X_t^\# \\
\Gamma[V_t] &= \text{delta}_t^2\, \Gamma[X_t] \\
\Gamma[V_s, V_t] &= \text{delta}_s \text{delta}_t\, \Gamma[X_s, X_t] \\
\Gamma[H_t] &= \text{gamma}_t^2\, \Gamma[X_t] \\
\Gamma[H_s, H_t] &= \text{gamma}_s \text{gamma}_t\, \Gamma[X_s, X_t] \\
\Gamma[V_s, H_t] &= \text{delta}_s \text{gamma}_t\, \Gamma[X_s, X_t] \\
\Gamma[X_t] &= M_t^2 \int_0^t \frac{\alpha(u)X_u^2\sigma^2(u, X_u)}{M_u^2}\,du \\
\Gamma[X_s, X_t] &= M_s M_t \int_0^{s\wedge t} \frac{\alpha(u)X_u^2\sigma^2(u, X_u)}{M_u^2}\,du.
\end{aligned}$$

Exercise (The so-called "feedback effect"). Let us return to the asset price model, i.e. to the equation

$$dX_t = X_t\sigma(t, X_t)\,dB_t + X_t r(t)\,dt.$$

We have already calculated $X_t^\#$ for the W.O.U. error structure. $X_t^\#$ is a semimartingale defined on the space $(\Omega, \mathcal{A}, \mathbb{P}) \times (\hat{\Omega}, \hat{\mathcal{A}}, \hat{\mathbb{P}})$ and by relation (44) we have

$$dX_t^\# = \big(\sigma(t, X_t) + X_t\sigma_x'(t, X_t)\big)X_t^\#\,dB_t + \sqrt{\alpha(t)}X_t\sigma(t, X_t)\,d\hat{B}_t + r(t)X_t^\#\,dt.$$

Let us suppose that the function σ is regular and does not vanish, then the process $X_t\sigma(t, X_t)$ is a semimartingale and we can apply Itô calculus:

$$\begin{aligned}
d\big(X_t\sigma(t, X_t)\big) &= X_t\sigma(t, X_t)\big(\sigma(t, X_t) + X_t\sigma_x'(t, X_t)\big)\,dB_t \\
&\quad + \Big[\big(\sigma(t, X_t) + X_t\sigma_x'(t, X_t)\big)X_t r(t) + \frac{1}{2}X_t^3\sigma^2(t, X_t)\sigma_{x^2}''(t, X_t) \\
&\quad + X_t^2\sigma^2(t, X_t)\sigma_x'(t, X_t) + X_t\sigma_t'(t, X_t)\Big]\,dt,
\end{aligned}$$

which gives, still by Itô calculus

$$d\frac{X_t^\#}{X_t\sigma(t,X_t)} = \sqrt{\alpha(t)}\,d\hat{B}_t - X_t^\# \frac{X_t^2\sigma_x'(t,X_t)r(t) + \frac{1}{2}X_t^3\sigma^2\sigma_{x^2}''(t,X_t) + X_t^2\sigma^2\sigma_x'(t,X_t) + X_t\sigma_t'(t,X_t)}{X_t^2\sigma^2(t,X_t)}dt.$$

Setting $Z_t = \frac{X_t^\#}{X_t\sigma(t,X_t)}$, we obtain

$$dZ_t = \sqrt{\alpha(t)}\,d\hat{B}_t + Z_t\lambda(t)\,dt$$

with

$$\lambda(t) = \left[-\frac{\sigma_x'(t,X_t)}{\sigma(t,X_t)}r(t) - \frac{\sigma(t,X_t)}{2}\big(X_t\sigma_{x^2}''(t,X_t) + 2\sigma_x'(t,X_t)\big) - \frac{\sigma_t'(t,X_t)}{X_t\sigma(t,X_t)}\right]X_t.$$

Hence

$$Z_t = R_t\int_0^t \frac{\sqrt{\alpha(s)}}{R_s}\,d\hat{B}_s \quad \text{with } R_t = \exp\int_0^t \lambda(s)\,ds. \tag{51}$$

In their study, Barucci, Malliavin et al. [2003] call the $\lambda(t)$ process the *feedback effect rate* and assign it the interpretation of a sensitivity of stock price to its own volatility.

Let us note herein that relation (51) immediately yields

$$\frac{\Gamma[X_t]}{X_t^2\sigma^2(t,X_t)} = R_t^2\int_0^t \frac{\alpha(s)}{R_s^2}\,ds. \tag{52}$$

We observe that the proportional quadratic error $\frac{\Gamma[X_t]}{X_t^2}$ (or equivalently the quadratic error on $\log(X_t)$ since $\Gamma\big[\log(X_t)\big] = \frac{\Gamma[X_t]}{X_t^2}$) when divided by the squared volatility is a *process with finite variation* (see (52)). In other words, it is a process whose randomness is relatively quiet.

This finding means that if the Brownian motion is perturbed in the following way

$$\omega(\cdot) \to \int_0^\cdot e^{-\frac{\alpha(s)}{2}\varepsilon}dB_s + \int_0^\cdot \sqrt{1-e^{-\alpha(s)\varepsilon}}\,d\hat{B}_s,$$

the stock price is perturbed such that the martingale part of the logarithm of its proportional error $\log\frac{\Gamma[X_t]}{X_t^2}$ is equal to the martingale part of $\log\sigma^2(t,X_t)$.

4.2 Sensitivity of the solution of an S.D.E. to errors on its functional coefficients. Let us recall the notation and main results of Chapter V, Section 4, where we explained how to equip a functional space with an error structure in order to study the sensitivity of a model to a functional parameter.

For the sake of simplicity, we suppose that the function f is from $\mathbb{R}$ into $\mathbb{R}$ and expands in series

$$f = \sum_n a_n \xi_n$$

with the ξ_n's being a basis of a vector space.

Choice of the probability measure. We randomize f by supposing the a_n's to be random and independent, but not identically distributed. We also suppose that the law μ_n of a_n can be written

$$\mu_n = \alpha_n \mu + (1 - \alpha_n)\delta_0,$$

with $\alpha_n \in]0, 1[$ and $\sum_n \alpha_n < +\infty$. Then (see Chapter V, Section 4) the probability measure $\mathbb{P}_1 = \bigotimes_n \mu_n$ is such that only a finite number of the a_n's are non zero, and the scaling $f \to \lambda f$ $(\lambda \neq 0)$ gives from $\mathbb{P}$ an absolutely continuous measure.

Error structure. Let us take an error structure

$$\big(\Omega_1, \mathcal{A}_1, \mathbb{P}_1, \mathbb{D}_1, \mathbb{P}_2\big) = \prod_n \big(\mathbb{R}, \mathcal{B}(\mathbb{R}), \mu_n, \boldsymbol{d}_n, \gamma_n\big)$$

with the a_n's being the coordinated mappings. We suppose that the error structures $\big(\mathbb{R}, \mathcal{B}(\mathbb{R}), \mu_n, \boldsymbol{d}_n, \gamma_n\big)$ are such that the identity map belongs to $\boldsymbol{d}_n$ and possess a sharp operator #. This provides by product a sharp operator on $\big(\Omega_1, \mathcal{A}_1, \mathbb{P}_1, \mathbb{D}_1, \mathbb{P}_1\big)$ (see the remark in Chapter V, Section 2.2, p. 80).

Error calculus. We now suppose that the functions ξ_n from $\mathbb{R}$ into $\mathbb{R}$ are of class $\mathcal{C}^1 \cap \mathrm{Lip}$. It follows that the function $f = \sum_n a_n \xi_n$ is $\mathbb{P}_1$-a.s. of class $\mathcal{C}^1 \cap \mathrm{Lip}$ (since the sum is almost surely finite).

Let X be an erroneous quantity with values in $\mathbb{R}$ defined on an error structure $\big(\Omega_2, \mathcal{A}_2, \mathbb{P}_1, \mathbb{D}_2, \Gamma_2\big)$ equipped with a sharp operator. Then on

$$(\Omega, \mathcal{A}, \mathbb{P}, \mathbb{D}, \Gamma) = \big(\Omega_1, \mathcal{A}_1, \mathbb{P}_1, \mathbb{D}_1, \Gamma_1\big) \times \big(\Omega_2, \mathcal{A}_2, \mathbb{P}_2, \mathbb{D}_2, \Gamma_2\big)$$

we have the following useful formula

$$\big(f(X)\big)^{\#} = f^{\#}(X) + f'(X)X^{\#} \tag{53}$$

which means

$$\Big(\sum_n a_n \xi_n(X)\Big)^{\#} = \sum_n a_n^{\#} \xi_n(X) + \sum_n a_n \xi_n'(X) X^{\#} \tag{54}$$

as long as the integrability condition guaranteeing that $f(X)$ belongs to $\mathbb{D}$ is fulfilled. Here this condition is

$$\iint \Big(\sum_n \Gamma_1[a_n]\xi_n^2(X) + \sum_{m,n} a_m a_n \xi_m'(X)\xi_n'(X)\Gamma_2[X]\Big) d\mathbb{P}_1 d\mathbb{P}_2 < +\infty.$$

With these hypotheses formula (53) is a direct application of the theorem on product error structures.

Remark. Let us emphasize that formula (53) is quite general and still remains valid if X and the error on X are correlated with f and the error on f. This clearly appears in formula (54). Only the integrability conditions are less simple.

Financial model. We now take the same model as before. The asset is modeled by the s.d.e.

$$dX_t = X_t\sigma(t, X_t)\,dB_t + X_t r(t)\,dt, \qquad X_0 = x.$$

The interest rate is deterministic and the probability is a martingale measure (see equations (41), (42) and (43) of Section 4.1).

4.2.1 Sensitivity to local volatility. An error is introduced on σ under hypotheses similar to what was recalled above, in such a way that the following formula is valid:

$$\big(\sigma(t, Y)\big)^{\#} = \sigma^{\#}(t, Y) + \sigma'_x(t, Y)Y^{\#}$$

where Y is a random variable, eventually correlated with σ, such that $\sigma(t, Y) \in \mathbb{D}$ (see the preceding remark).

From the equation

$$X_t = x + \int_0^t X_s\sigma(s, X_s)\,dB_s + \int_0^t X_s r(s)\,ds$$

we have

$$X_t^{\#} = \int_0^t \big(X_s^{\#}\sigma(s, X_s) + X_s\sigma^{\#}(s, X_s) + X_s\sigma'_x(s, X_s)X_s^{\#}\big)\,dB_s + \int_0^t X_s^{\#} r(s)\,ds. \tag{55}$$

We then set

$$K_s = \sigma(s, X_s) + X_s\sigma'_x(s, X_s)$$

and

$$M_t = \exp\left\{\int_0^t K_s\,dB_s - \frac{1}{2}\int_0^t K_s^2\,ds + \int_0^t r(s)\,ds\right\}.$$

Equation (55) then has the following solution:

$$X_t^{\#} = M_t \int_0^t \frac{X_s\sigma^{\#}(s, X_s)}{M_s}\big(dB_s - K_s\,ds\big). \tag{56}$$

We will extend further the calculations in three particular cases.

First case. $\sigma(t, x)$ is represented on the basis of a vector space consisting of function $\psi_n(t, x)$ regular in x. We set

$$\sigma(t, x) = \sum_n a_n\psi_n(t, x)$$

and follow the approach sketched out at the beginning of Section 4.2 with

$$\begin{aligned}\Gamma[a_n] &= a_n^2\\ \Gamma[a_m, a_n] &= 0 \quad \text{for } m \neq n\\ a_n^{\#} &= a_n \frac{\hat{a}_n - \hat{\mathbb{E}}\hat{a}_n}{\beta_n} \qquad \beta_n = \sqrt{\hat{\mathbb{E}}(\hat{a}_n - \hat{\mathbb{E}}\hat{a}_n)^2}.\end{aligned}$$

Thus

$$\sigma^{\#}(s, X_s) = \sum_n a_n^{\#} \psi_n(s, X_s)$$

$$X_t^{\#} = \sum_n M_t \int_0^t \frac{X_s \psi_n(s, X_s)}{M_s} (dB_s - K_s\, ds) a_n^{\#} \tag{57}$$

and

$$\Gamma[X_t] = \sum_n M_t^2 \left(\int_0^t \frac{X_s \psi_n(s, X_s)}{M_s} (dB_s - K_s\, ds) \right)^2 a_n^2. \tag{58}$$

In order to obtain the error on the value of a European option, we start with formula (42) of Section 4.1:

$$V_t = \exp\left(- \int_t^T r(s)\, ds \right) \mathbb{E}\left[f(X_T) \mid \mathcal{F}_t \right]$$

which yields

$$\begin{aligned} V_t^{\#} &= \exp\left(- \int_t^T r(s)\, ds \right) \left(\mathbb{E}\left[f(X_T) \mid \mathcal{F}_t \right] \right)^{\#} \\ &= \exp\left(- \int_t^T r(s)\, ds \right) \mathbb{E}\left[(f(X_T))^{\#} \mid \mathcal{F}_t \right] \end{aligned}$$

as can be seen by writing

$$\mathbb{E}[Z \mid \mathcal{F}_t] = \tilde{\mathbb{E}}[Z(w, \tilde{w})]$$

with

$$\begin{aligned} w &= (s \to \omega(s), s \leq t) \\ \tilde{w} &= (s \to \omega(s) - \omega(t), s \geq t). \end{aligned}$$

Hence

$$V_t^{\#} = \exp\left(- \int_t^T r(s)\, ds \right) \mathbb{E}\left[f'(X_T) X_T^{\#} \mid \mathcal{F}_t \right]$$

i.e. from (57)

$$
\begin{aligned}
V_t^{\#} &= \exp(-\int_t^T r(s)\,ds) \\
&\quad \times \sum_n \mathbb{E}\Big[f'(X_T)M_T\int_0^T \frac{X_s\psi_n(s,X_s)}{M_s}(dB_s - K_s\,ds) \mid \mathcal{F}_t\Big]a_n^{\#}.
\end{aligned} \tag{59}
$$

If we set

$$
V_t^n = \exp\Big(-\int_t^T r(s)\,ds\Big)\,\mathbb{E}\Big[f'(X_T)M_T\int_0^T \frac{X_s\psi_n(s,X_s)}{M_s}(dB_s - K_s\,ds) \mid \mathcal{F}_t\Big]
$$

which is the value of a European option of payoff

$$
f'(X_T)M_T\int_0^T \frac{X_s\psi_n(s,X_s)}{M_s}\big(dB_s - K_s\,ds\big)
$$

we observe that (59) gives

$$
\Gamma\big[V_t\big] = \sum_n \big(V_t^n\big)^2 a_n^2.
$$

Remark. If $X_{t,n}^{\varepsilon}$ is the solution to

$$
\begin{cases} dX_t^{\varepsilon} = X_t^{\varepsilon}\big(\sigma(t,X_t^{\varepsilon}) + \varepsilon\psi_n(t,X_t^{\varepsilon})\big)\,dB_t + X_t^{\varepsilon}r(t)\,dt \\ X_0^{\varepsilon} = x \end{cases}
$$

and if the corresponding value of the option is $V_{t,n}^{\varepsilon}$, we have

$$
V_t^n = \frac{\partial V_{t,n}^{\varepsilon}}{\partial\varepsilon}\Big|_{\varepsilon=0}.
$$

◇

In order to obtain the error on the hedging portfolio, we start with expression (49) of Section 4.1

$$
H_t = \exp\Big(-\int_t^T r(s)\,ds\Big)\,\mathbb{E}\Big[f'(X_T)\frac{M_T}{M_t} \mid \mathcal{F}_t\Big].
$$

We have

$$
\begin{aligned}
H_t^{\#} = \exp\Big(-\int_t^T r(s)\,ds\Big)\,\mathbb{E}\Big\{\frac{M_T}{M_t}\Big[&f''(X_T)M_T\int_0^T \frac{X_s\sigma^{\#}(s,X_s)}{M_s}(dB_s - K_s\,ds) \\
&+ f'(X_T)\int_t^T K_s^{\#}(dB_s - K_s\,ds)\Big] \mid \mathcal{F}_t\Big\}.
\end{aligned}
$$

The calculation can be extended and will eventually give a linear expression in $a_n^\#$. Hence $\Gamma[H_t]$ will be of the form

$$\Gamma\big[H_t\big] = \sum_n \big(h_t^n\big)^2 a_n^2. \tag{60}$$

Second case. We suppose here the volatility to be local and *stochastic*

$$\sigma(t, y, w)$$

and given by a diffusion independent of $(B_t)_{t\geq 0}$. In other words

$$\sigma(t, y, w) = \sigma_t^y(w),$$

where σ_t^y is the solution to

$$\begin{cases} d\sigma_t = a(\sigma_t)\, dW_t + b(\sigma_t)\, dt \\ \sigma_0 = c(y) \end{cases}$$

with $(W_t)_{t\geq 0}$ a Brownian motion independent of $(B_t)_{t\geq 0}$.

If functions a, b and c are regular, the mapping $y \to \sigma(t, y, w)$ is regular and we suppose that the formula

$$\big(\sigma(t, Y)\big)^\# = \sigma^\#(t, Y) + \sigma'_y(t, Y)Y^\#$$

is valid in the following calculation.

On $(W_t)_{t\geq 0}$ we introduce an error of the Ornstein–Uhlenbeck type. Setting

$$m_t^y = \exp\left\{\int_0^t a'(\sigma_s^y)\, dW_s - \frac{1}{2}\int_0^t a'^2(\sigma_s^y)\, ds + \int_0^t b'(\sigma_s^y)\, ds\right\},$$

a previously conducted calculation yields

$$\sigma^\#(t, y) = c'(y) m_t^y \int_0^t \frac{a(\sigma_s^y)}{m_s^y}\, d\hat{W}_s.$$

We then follow the calculation of the first case with this $\sigma^\#$:

$$X_t^\# = M_t \int_0^t \frac{X_s c'(X_s) m_s^{X_s}}{M_s} \int_0^s \frac{a\big(\sigma_u^{X_s}\big)}{m_u^{X_s}}\, d\hat{W}_u \big(dB_s - K_s\, ds\big).$$

We place outside the integral with respect to $d\hat{W}_u$

$$X_t^\# = M_t \int_0^t \int_u^t \frac{X_s c'(X_s) m_s^{X_s} a\big(\sigma_u^{X_s}\big)}{M_s m_u^{X_s}} \big(dB_s - K_s\, ds\big)\, d\hat{W}_u$$

which gives

$$\Gamma\big[X_t\big] = M_t^2 \int_0^t \left(\int_u^t \frac{X_s c'(X_s) m_s^{X_s} a\big(\sigma_u^{X_s}\big)}{M_s m_u^{X_s}} (dB_s - K_s\, ds) \right)^2 du. \tag{61}$$

The calculation of $\Gamma\big[V_t\big]$ and $\Gamma\big[H_t\big]$ can be performed in a similar manner and yields expressions computable by Monte Carlo methods.

Third case. We suppose here that the local volatility is stochastic, with $\sigma(t, y)$ being a stationary process independent of $\big(B_t\big)_{t\geq 0}$. We idea involved is best explained in an example.

Consider k regular functions $\eta_1(y), \dots, \eta_k(y)$ and set

$$\sigma(t, y) = \sigma_0 e^{Y(t,y)}$$

with

$$Y(t, y) = \sum_{i=1}^{k} Z_i(t)\eta_i(y)$$

where $Z(t) = \big(Z_1(t), \dots, Z_k(t)\big)$ is an $\mathbb{R}^k$-valued stationary process.

For instance, in order to obtain real processes, we may take

$$Z_i(t) = \sum_j \int_0^\infty \xi_{ij}(\lambda)\big(\cos\lambda t\, dU_\lambda^j + \sin\lambda t\, dV_\lambda^j\big)$$

where $\xi_{ij} \in L^2(\mathbb{R}_+)$ and $U_\lambda^1, \dots, U_\lambda^k, V_j^1, \dots, V_\lambda^k$ are independent Brownian motions.

We introduce an error of the Ornstein–Uhlenbeck type on these Brownian motions characterized by the relation

$$\begin{aligned}\Gamma\bigg[\int_0^\infty f_1(\lambda)\, dU_\lambda^1 + \dots + \int_0^\infty f_k(\lambda)\, dU_\lambda^k \\ + \int_0^\infty g_1(\lambda)\, dV_\lambda^1 + \dots + \int_0^\infty g_k(\lambda)\, dV_\lambda^k\bigg] \\ = \int_0^\infty \Big(\sum_i f_i^2(\lambda) + \sum_j g_j^2(\lambda)\Big) d\lambda.\end{aligned}$$

The corresponding sharp operator is given by

$$\left(\int f_i(\lambda)\, dU_\lambda^i\right)^{\#} = \int f_i(\lambda)\, d\hat{U}_\lambda^i$$

and similarly for the V_λ^i's, so that

$$\big(Z_i(t)\big)^{\#} = \widehat{Z_i(t)}$$

where $\widehat{Z_i(t)}$ denotes the process $Z_i(t)$ constructed with $\hat{U}^1, \dots, \hat{U}^k, \hat{V}^1, \dots, \hat{V}^k$. Thus

$$Y^\#(t, y) = \hat{Y}(t, y)$$

and

$$\sigma^\#(t, y) = \sigma(t, y)\hat{Y}(t, y).$$

We now follow the calculation with this $\sigma^\#$:

$$X_t^\# = M_t \int_0^t \frac{X_s \sigma(s, X_s)\hat{Y}(s, X_s)}{M_s}\big(dB_s - K_s\, ds\big).$$

Since $\hat{Y}(s, y)$ is linear in $\hat{U}^1, \dots, \hat{U}^k, \hat{V}^1, \dots, \hat{V}^k$, this yields for $\Gamma[X_t]$ a sum of squares:

$$\begin{aligned}\Gamma\big[X_t\big] = \int_0^\infty M_t^2 \sum_{ij}\Bigg[&\bigg(\int_0^t \frac{X_s\sigma(s, X_s)}{M_s} y_i(X_s)\xi_{ij}(\lambda)\cos\lambda s\big(dB_s - K_s\, ds\big)\bigg)^2 \\ &+ \bigg(\int_0^t \frac{X_s\sigma(s, X_s)}{M_s} y_i(X_s)\xi_{ij}(\lambda)\sin\lambda s\big(dB_s - K_s\, ds\big)\bigg)^2\Bigg]d\lambda.\end{aligned} \tag{62}$$

Comment. We have developed these sensitivity calculations to functional σ mainly in order to highlight the extend of the language of error structures. In numerical applications, a sequence of perturbations of σ in fixed directions (à la Gateaux) will sometimes be sufficient, but the formulae with series expansions allow choosing the rank at which the series are to be truncated.

In addition, the framework of error structures guarantees that the calculation can be performed with functions of class $\mathcal{C}^1 \cap \mathrm{Lip}$.

4.2.2 Sensitivity to the interest rate. The calculations of interest rate sensitivity are simpler since there are fewer terms due to Itô's formula. We suppose $r(t)$ to be represented as

$$r(t) = \sum_n b_n \rho_n(t)$$

where the functions $\rho_n(t)$ are deterministic and the b_n's satisfy hypotheses similar to the preceding a_n's:

$$\begin{aligned}\Gamma\big[b_n\big] &= b_n^2 \\ \Gamma\big[b_m, b_n\big] &= 0 \quad \text{for } m \neq n \\ b_n^\# &= b_n \frac{\hat{b}_n - \hat{\mathbb{E}}\hat{b}_n}{\sqrt{\hat{\mathbb{E}}\big(\hat{b}_n - \hat{\mathbb{E}}\hat{b}_n\big)^2}}\end{aligned}$$

such that
$$r^{\#}(t) = \sum_n b_n^{\#} \rho_n(t).$$

From the equation
$$X_t = x + \int_0^t X_s \sigma(s, X_s)\, dB_s + \int_0^t X_s r(s)\, ds$$

we obtain
$$X_t^{\#} = \int_0^t K_s X_s^{\#}\, dB_s + \int_0^t \big(X_s^{\#} r(s) + X_s r^{\#}(s)\big)\, ds$$

and
$$X_t^{\#} = M_t \int_0^t \frac{X_s r^{\#}(s)}{M_s}\, ds, \tag{63}$$

where M_s and K_s have the same meaning as before. It then follows that
$$\Gamma[X_t] = M_t^2 \bigg(\sum_n \int_0^t \frac{X_s \rho_n(s)}{M_s}\, ds \bigg)^2 b_n^2. \tag{64}$$

The error on the value of the option is obtained from the formula
$$V_t = \exp\bigg(- \int_t^T r(s)\, ds \bigg) \mathbb{E}[f(X_T) \mid \mathcal{F}_t]$$

and
$$\begin{aligned} V_t^{\#} = \exp\bigg(- \int_t^T r(s)\, ds \bigg) \mathbb{E}\bigg[- \int_t^T r^{\#}(s)\, ds f(X_T) \\ + f'(X_T) M_T \int_0^T \frac{X_s r^{\#}(s)}{M_s}\, ds \mid \mathcal{F}_t \bigg] \end{aligned} \tag{65}$$

yields
$$\begin{aligned} \Gamma[V_t] = \exp\bigg(-2 \int_t^T r(s)\, ds \bigg) \sum_n \bigg(\mathbb{E}\bigg[- \int_t^T \rho_n(s)\, ds f(X_T) \\ + f'(X_T) M_T \int_0^T \frac{X_s \rho_n(s)}{M_s}\, ds \mid \mathcal{F}_t \bigg] \bigg)^2 b_n^2. \end{aligned} \tag{66}$$

4.3 Error on local volatility due solely to the trader. We now suppose that the model
$$X_t = x + \int_0^t X_s \sigma(s, X_s)\, dB_s + \int_0^t X_s r(s)\, ds$$

is accurate and contains no error.

To manage a European option of payoff $f(X_T)$ we suppose, as in Section 3.2.2, that the trader has performed a correct initial pricing, but that from 0 to T his hedging portfolio is incorrect due to an error on the function σ.

The hedging equation is

$$\tilde{V}_t = V_0 + \int_0^T H_s \, d\tilde{X}_s \tag{67}$$

where

$$\tilde{V}_t = \exp\left(-\int_0^t r(s)\,ds\right) V_t$$

and

$$\tilde{X}_t = \exp\left(-\int_0^t r(s)\,ds\right) X_t$$

with

$$V_t = \exp\left(-\int_t^T r(s)\,ds\right) \mathbb{E}\big[f(X_T) \mid \mathcal{F}_t\big]$$

$$H_t = \exp\left(-\int_t^T r(s)\,ds\right) \mathbb{E}\big[f'(X_T)M_T \mid \mathcal{F}_t\big]\frac{1}{M_t}. \tag{68}$$

By committing an error on H_t, the trader does not realize at exercise time T the discounted payoff $\tilde{V} = \exp\big(-\int_0^T r(s)\,ds\big) f(X_T)$ but rather

$$\tilde{P}_T = V_0 + \int_0^T \big[H_s\big] \, d\tilde{X}_s \tag{69}$$

where H_s is calculated using (68) with an incorrect function σ what we denote by brackets $\big[H_s\big]$. We must emphasize herein a very important observation.

Remark. In the case of a probabilistic model involving the parameter λ (here the volatility), if we were to consider this parameter erroneous by defining it on an error structure, all model quantities become erroneous. If, by means of mathematical relations of the model, a quantity can be written identically in two ways

$$X = \varphi(\omega, \lambda)$$
$$X = \psi(\omega, \lambda),$$

then the error on X calculated by φ or ψ will be the same.

Let us now suppose that the model is used for a decision thanks to a formula of practical utility, and that in this formula the user chooses the wrong value of λ *in certain places* where λ occurs but not everywhere, then the error may depend on the algebraic form of the relation used.

Let us consider a simple example.

Let L be the length of the projection of a triangle with edges of lengths a_1, a_2, a_3 and polar angles

$$\theta_1, \quad \theta_2, \quad \theta_3.$$

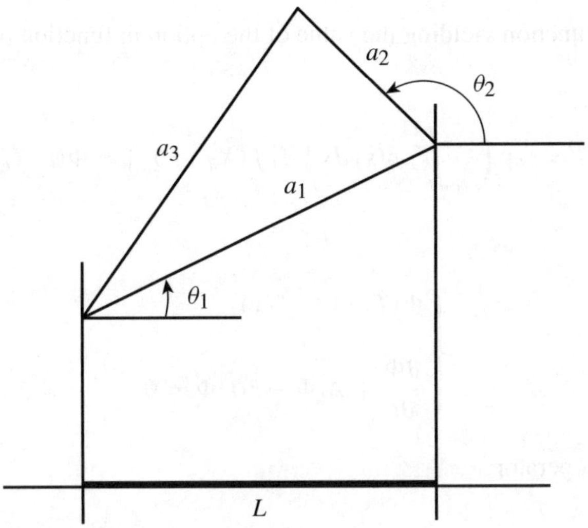

The length L satisfies

$$L = \max_{i=1,2,3} |a_i \cos(\theta_i)|$$

$$L = \frac{1}{2} \sum_{i=1,2,3} |a_i \cos(\theta_i)|.$$

If the user makes a mistake on the length a_1, and only on a_1 (without any attempt to respect the triangle by changing the other quantities) then the error on L will depend on use of the first or second formula. (Since the first formula gives a nonzero error only when the term $|a_1 \cos(\theta_1)|$ dominates the others.) ◇

In our problem, we must therefore completely specify both the formula used by the trader and where his use of σ is erroneous. Formula (68) is not sufficiently precise. The trader's action consists of using the Markov character of X_t in order to write the conditional expectation in the form

$$\mathbb{E}\left[f'(X_T)\frac{M_T}{M_t} \mid \mathcal{F}_t\right] = \Psi(t, X_t).$$

When computing Ψ, he introduces an error on σ and he is correct on X_t, which is given by the market (since we have assumed the asset price model to be accurate).

Let us recall some formulae given by classical financial theory (see Lamberton–Lapeyre, (1995), Chapter 5).

The function Ψ is given by

$$\Psi(t,x) = \frac{\partial \Phi}{\partial x}(t,x) \tag{70}$$

where Φ is the function yielding the value of the option in function of the stock price X_t:

$$V_t = \exp\left(-\int_t^T r(s)\,ds\right)\mathbb{E}\big[f(X_T) \mid \mathcal{F}_t\big] = \Phi(t, X_t). \tag{71}$$

It satisfies

$$\begin{cases} \Phi(T,x) = f(x) \\ \dfrac{\partial \Phi}{\partial t} + A_t\Phi - r(t)\Phi = 0 \end{cases} \tag{72}$$

where A_t is the operator

$$A_t u(x) = \frac{1}{2}x^2\sigma^2(t,x)\frac{\partial^2 u}{\partial x^2}(x) + xr(s)\frac{\partial u}{\partial x}(x).$$

We are interested in calculating

$$(\tilde{P}_T)^{\#} = \int_0^T \Psi^{\#}(t, X_t)\,d\tilde{X}_t = \int_0^T \frac{\partial \Phi^{\#}}{\partial x}(t, X_t)\,d\tilde{X}_t. \tag{73}$$

Taking (72) into account the (random) function $\Phi^{\#}(t,x)$ satisfies

$$\begin{cases} \Phi^{\#}(T,x) = 0 \\ \dfrac{\partial \Phi^{\#}}{\partial t} + A_t\Phi^{\#} + A_t^{\#}\Phi - r(t)\Phi^{\#} = 0 \end{cases} \tag{74}$$

where $A_t^{\#}$ is the operator

$$A_t^{\#}u(x) = \frac{1}{2}x\sigma(t,x)\sigma^{\#}(t,x)\frac{\partial^2 u}{\partial x^2}(x).$$

In order to calculate (73) let us apply the Itô formula to

$$\exp\left(-\int_0^t r(s)\,ds\right)\Phi^{\#}(t, X_t).$$

In conjunction with (74) this yields

$$\begin{aligned}
&\exp\Big(-\int_0^t r(s)\,ds\Big)\Phi^\#(t,X_t)\\
&\quad= \Phi^\#(0,X_0) + \int_0^t \exp\Big(-\int_0^s r(u)\,du\Big)\frac{\partial\Phi^\#}{\partial x}(s,X_s)\,dX_s\\
&\qquad - \int_0^t \exp\Big(-\int_0^s r(u)\,du\Big)X_s r(s)\frac{\partial\Phi^\#}{\partial x}(s,X_s)\,ds\\
&\qquad - \int_0^t \exp\Big(-\int_0^s r(u)\,du\Big)(A_s^\#\Phi)(s,X_s)\,ds.
\end{aligned} \tag{75}$$

Then, in introducing $\tilde{X}_s$ and setting $t = T$,

$$\int_0^t \frac{\partial\Phi^\#}{\partial x}(s,X_s)\,d\tilde{X}_s = -\Phi^\#(0,X_0) + \int_0^t \exp\left(-\int_0^s r(u)\,du\right)\left(A_s^\#\Phi\right)(s,X_s)\,ds.$$

Hence we have obtained

$$(\tilde{P}_T)^\# = -\Phi^\#(0,X_0) + \frac{1}{2}\int_0^T \tilde{X}_s\sigma(s,X_s)\sigma^\#(s,X_s)\frac{\partial^2\Phi}{\partial x^2}(s,X_s)\,ds. \tag{76}$$

Comment. Let us first note that we have not yet specified $\sigma^\#$. The preceding calculation is, for the time being, valid when the error committed by the trader is modeled as in any of the three cases discussed above.

First case.

$$\sigma(t,y) = \sum_n a_n\psi_n(t,y)$$
$$\sigma^\#(t,y) = \sum_n a_n^\#\psi_n(t,y).$$

Second case. σ is an independent diffusion

$$\sigma^\#(t,y) = c'(y)m_t^y\int_0^t \frac{a(\sigma_s^y)}{m_s^y}\,d\hat{W}_s.$$

Third case. σ is an independent stationary process

$$\sigma(t,x) = \sigma_0\exp\big(Y(t,y)\big)$$
$$\sigma^\#(t,x) = \sigma(t,x)\hat{Y}(t,y).$$

The first term in formula (76): $-\Phi^\#(0,X_0)$ stems from the fact that the trader is accurate on the pricing, hence his pricing is not all that coherent with the stochastic integral he uses for hedging. $\Phi^{\#}(0, X_0)$ can be interpreted as the difference between the pricing the trader would have proposed and that of the market (i.e. that of the model).

In the second term

$$\frac{1}{2}\int_0^T \tilde{X}_s \sigma(s, X_s)\sigma^{\#}(s, X_s)\frac{\partial^2 \Phi}{\partial x^2}(s, X_s)\,ds$$

the quantity $\sigma^{\#}(t, x)$ is a random derivative in the sense of Dirichlet forms. In several cases, it can be interpreted in terms of directional derivatives (see Bouleau–Hirsch [1991], Chapter II, Section 4). We can then conclude that if the payoff is a (regular and) convex function of asset price and if $\sigma(t, x)$ possesses a positive directional derivative in the direction of $y(t, x)$, this second term is positive. In other words, if the trader hedges with a function σ distorted in the direction of such a function y, his final loss is lower than the difference between the pricing he would have proposed and that of the market, since the second term is positive. We rediscover here some of the results in [El Karoui et al. 1998].

Comments on this chapter. We did not address the issue of American options, which are obviously more difficult to handle mathematically. Both the approach by instantaneous error structures and the study of the sensitivity of a model to its parameters or to a mistake of the trader are more subtle. One approach to this situation would be to choose hypotheses such that good convergence results of time discretization quantities are available. Thus, if V_t^n were the value of the option in a time discretization of step $\frac{T}{n}$, the calculation of $\Gamma\big[V_t^n\big]$ should be possible since the max and inf operators over a finite number of quantities are Lipschitz operations. If the hypotheses allow to show that $V_t^n \xrightarrow[n\uparrow\infty]{} V_t$ in $L^2(\mathbb{P})$ and $\Gamma\big[V_t^m - V_t^n\big] \xrightarrow[m,n,\uparrow\infty]{} 0$ in $L^1(\mathbb{P})$, then the closedness of the Dirichlet form will yield the conclusion that $\Gamma[V_t^n] \xrightarrow[n\uparrow\infty]{} \Gamma[V_t]$ in $L^1(\mathbb{P})$.

On the other hand, we did not calculate the biases neither for the Black–Scholes model, nor for the diffusion model, nor when the error is due solely to the trader. These omissions are strictly for the sake of brevity of this course since the appropriate tools were indeed available. Let us nevertheless mention that for the Black–Scholes model, an error on Brownian motion of the Ornstein–Uhlenbeck type provides the following biases:

$$\begin{aligned} A[S_t] &= -S_t\sigma B_t + \frac{1}{2}\sigma^2 S_t t \\ A[V_t] &= \text{delta}_t A[S_t] + \frac{1}{2}\,\text{gamma}_t\,\Gamma[S_t] \end{aligned}$$

(where $\Gamma[S_t] = S_t^2\sigma^2 t$)

$$A[H_t] = \text{gamma}_t A[S_t] + \frac{1}{2}\frac{\partial^3 F}{\partial x^3}(t, S_t, \sigma, r)\Gamma[S_t].$$

In the following chapter, which concerns a number of physics applications , we will provide examples in which the biases modify the experimental results.

Bibliography for Chapter VII

Books

N. Bouleau, *Martingales and Financial Markets*, Springer-Verlag, 2003.

N. Bouleau and F. Hirsch, *Dirichlet Forms and Analysis on Wiener Space*, Walter de Gruyter, 1991.

Cl. Dellacherie and P. A. Meyer, *Probabilités et Potentiel*, Hermann, 1987 (especially Chapter XIII, Construction de résolvantes et de semi-groupes).

M. Fukushima, Y. Oshima and M. Takeda, *Dirichlet Forms and Markov Processes*, Walter de Gruyter, 1994.

D. Lamberton and B. Lapeyre, *Introduction to Stochastic Calculus Applied to Finance*, Chapman & Hall, London, 1995.

Z.-M. Ma and M. Röckner, *Introduction to the Theory of (Non-symmetric) Dirichlet Forms*, Springer-Verlag, 1992.

K. Yosida, *Functional Analysis*, Springer-Verlag, fourth ed., 1974 (especially Chapter IX).

Articles

E. Barucci, P. Malliavin, M. E. Mancino, R. Renó and A. Thalmaier, The price-volatility feedback rate: an implementable mathematical indicator of market stability, *Math. Finance* **13** (2003), 17–35.

Ch. Berg and G. Forst, Non-symmetric translation invariant Dirichlet forms, *Invent. Math.* **21** (1973), 199–212.

H.-P. Bermin and A. Kohatsu-Higa, Local volatility changes in the Black–Scholes model, *Economics Working Papers*, Univ. Pompeu Fabra, Sept. 1999.

N. Bouleau, Error calculus and path sensitivity in financial models, *Math. Finance* **13** (2003), 115–134.

C. Constantinescu, N. El Karoui and E. Gobet, Représentation de Feynman-Kac dans des domainestemps-espaces et sensibilité par rapport au domaine, *C.R. Acad. Sci. Paris Sér. I* **337** (2003).

N. El Karoui, M. Jeanblanc-Picqué and St. Shreve, Robustness of the Black and Scholes formula, *Math. Finance* **8** (1998), 93–126.

J. P. Fouque and T. A. Tullie, variance reduction for Monte-Carlo simulation in a stochastic volatility environment, preprint, 2001.

S. L. Heston, A closed-form solution for options with stochastic volatility with applications to bond and currency options, *Rev. Financial Stud.* **6** (1993), 327–343.

J. Hull and A. White, The pricing of options on assets with stochastic volatilities, *J. Finance* **42** (1987), 281–300;
An analysis of the bias in option pricing caused by a stochastic volatility, *Adv. Futures and Options Research* **3** (1988), 29–61.

P. A. Meyer, L'existence de $[M, M]$ et l'intégrale de Stratonovitch. *In* Un cours sur les intégrales stochastiques, *Sém. de Probabilités X*, Lecture Notes in Math. 511, Springer-Verlag, 1976.

L. O. Scott, Option pricing when the variance changes randomly: Theory, estimation and applications, *J. Financial and Quantitative Analysis* **22** (1987), 419–438.

E. M. Stein and J. C. Stein, Stock price distributions with stochastic volatility: An analytic approach, *Rev. Financial Studies* **4** (1991), 727–752.

J. B. Wiggins, Options values under stochastic volatility. Theory and empirical estimates, *J. Financial Economics* **19** (1987), 351–372.

Chapter VIII

Applications in the field of physics

Our aim in this chapter is not to propose explanations of phenomena or experimental results, but more modestly to provide examples of physical situations in which the language of error structures may be used.

The first four sections comprise exercises and discussion in order to lend some insight into using the language of error structures in a physical context.

The fifth section concerns the nonlinear oscillator submitted to thermal agitation and presents an example in which the lack of accuracy due to temperature acting upon a system may be modeled by an error structure. The nonlinearity of the oscillator gives rise to biases and these biases explain the expansion of crystals as temperature increases.

The sixth section yields mathematical arguments to support that some systems may be "naturally" provided with *a priori* error structures. In this study we extend the "arbitrary functions method" of Poincaré and Hopf from the case of probability theory to the case of error structures.

1 Drawing an ellipse (exercise)

An ellipse is drawn by placing the two foci at distance $2c$ with a graduated rule and then by using a string of length $2\ell > 2c$. We suppose that the pair (c, ℓ) is erroneous: it is defined as the coordinate mappings of the following error structure $(\Omega, \mathfrak{A}, \mathbb{P}, \mathbb{D}, \Gamma)$, where

$$\begin{aligned}
\Omega &= \left\{(x, y) \in \mathbb{R}_+^2 : y > x\right\} \\
\mathfrak{A} &= \mathcal{B}(\Omega) \\
\mathbb{P}(dx, dy) &= 1_\Omega \cdot e^{-y}\, dx\, dy \\
\Gamma[u](x, y) &= \alpha^2 {u'_x}^2(x, y) + \beta^2 {u'_y}^2(x, y), \quad \forall u \in \mathcal{C}_K(\Omega).
\end{aligned}$$

The closability of the associated form has been derived in Fukushima [1980], Chapter 2, Section 2.1 or Fukushima et al. [1994], Chapter 3, Section 3.1, which is similar to Example III.2.3 in the case of an absolutely continuous measure.

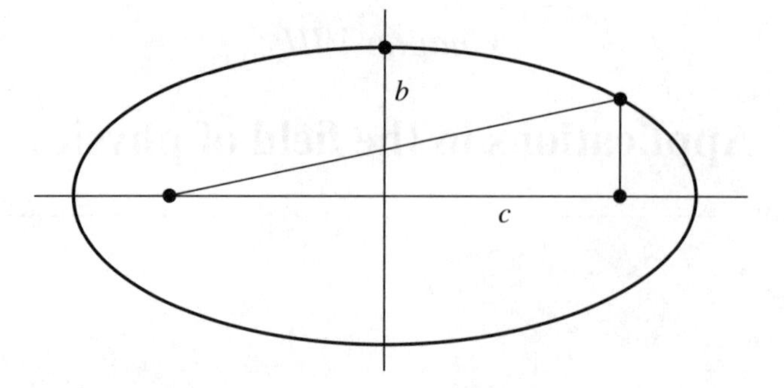

a) Error on the area of the ellipse. Let a be the major half-axis and b be the minor half-axis. From the formula

$$A = \pi ab = \pi\ell\sqrt{\ell^2 - c^2},$$

we obtain

$$A^\# = \pi\,\frac{-\ell c}{\sqrt{\ell^2 - c^2}}\,c^\# + \pi\,\frac{2\ell^2 - c^2}{\sqrt{\ell^2 - c^2}}\,\ell^\#$$

and

$$\Gamma[A] = \frac{\pi^2}{\ell^2 - c^2}\big[\ell^2 c^2 \alpha^2 + (2\ell^2 - c^2)^2 \beta^2\big].$$

b) Error on the length of the ellipse. The length is calculated by parametric representation

$$L = \int_0^{2\pi} \sqrt{\dot{X}^2(t) + \dot{Y}^2(t)}\,dt$$

with $X(t) = a\cos t$, $Y(t) = b\sin t$. This yields

$$L = \int_0^{2\pi} \sqrt{\ell^2 - c^2\cos^2 t}\,dt$$

$$L^\# = \int_0^{2\pi} \frac{\ell\ell^\# - cc^\#\cos^2 t}{\sqrt{\ell^2 - c^2\cos^2 t}}\,dt$$

and

$$\Gamma[L] = \alpha^2\left(\int_0^{2\pi} \frac{c\cos^2 t}{\sqrt{\ell^2 - c^2\cos^2 t}}\,dt\right)^2 + \beta^2\left(\int_0^{2\pi} \frac{\ell\,dt}{\sqrt{\ell^2 - c^2\cos^2 t}}\right)^2.$$

These classical elliptical integrals are tabulated. We also derive the correlation between the error on the area and the error on the length:

$$\Gamma[A, L] = \pi\alpha^2 \frac{\ell c}{\sqrt{\ell^2 - c^2}} \int_0^{2\pi} \frac{c\cos^2 t}{\sqrt{\ell^2 - c^2\cos^2 t}}\,dt$$
$$+ \pi\beta^2 \frac{2\ell^2 - c^2}{\sqrt{\ell^2 - c^2}} \int_0^{\pi} \frac{\ell\,dt}{\sqrt{\ell^2 - c^2\cos^2 t}}.$$

c) Errors on sectors. Let us consider two angular sectors M_1OM_1' and M_2OM_2' swept by the vector radius as t varies respectively from t_1 to t_1' and t_2 to t_2'. The areas are given by

$$A\big(M_1OM_1'\big) = \frac{b}{a}a\big(t_1' - t_1\big) = \sqrt{\ell^2 - c^2}\,\big(t_1' - t_1\big)$$
$$A\big(M_2OM_2'\big) = \sqrt{\ell^2 - c^2}\big(t_2' - t_2\big).$$

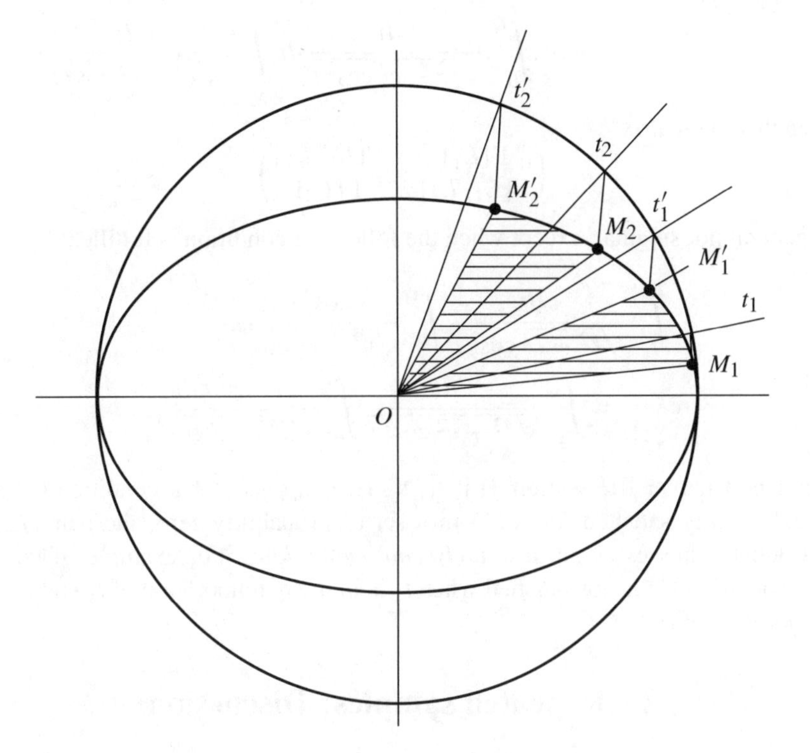

Let A_1 and A_2 be these two areas, we can then see that the matrix

$$\begin{pmatrix} \Gamma[A_1] & \Gamma[A_1, A_2] \\ \Gamma[A_1, A_2] & \Gamma[A_2] \end{pmatrix}$$

is singular since A_1 and A_2 are both functions of the single random variable $\sqrt{\ell^2 - c^2}$. The law of the pair $\big(A_1, A_2\big)$ is carried by a straight line of the plane.

For the arc lengths of these sectors we obtain

$$L_1 = \int_{t_1}^{t_1'} \sqrt{\ell^2 - c^2\cos^2 t}\,dt, \quad L_2 = \int_{t_2}^{t_2'} \sqrt{\ell^2 - c^2\cos^2 t}\,dt,$$

hence for $i = 1, 2$,

$$\Gamma[L_i] = \alpha^2 \left(\int_{t_i}^{t_i'} \frac{c \cos^2 t}{\sqrt{\ell^2 - c^2 \cos^2 t}} \, dt \right)^2 + \beta^2 \left(\int_{t_i}^{t_i'} \frac{\ell \, dt}{\sqrt{\ell^2 - c^2 \cos^2 t}} \right)^2$$

and

$$\begin{aligned}\Gamma[L_1, L_2] = \alpha^2 & \int_{t_1}^{t_1'} \frac{c \cos^2 t}{\sqrt{\ell^2 - c^2 \cos^2 t}} \, dt \int_{t_2}^{t_2'} \frac{c \cos^2 t}{\sqrt{\ell^2 - c^2 \cos^2 t}} \, dt \\ & + \beta^2 \int_{t_1}^{t_1'} \frac{\ell \, dt}{\sqrt{\ell^2 - c^2 \cos^2 t}} \, dt \int_{t_2}^{t_2'} \frac{\ell \, dt}{\sqrt{\ell^2 - c^2 \cos^2 t}}.\end{aligned}$$

We see that the matrix

$$\begin{pmatrix} \Gamma[L_1] & \Gamma[L_1, L_2] \\ \Gamma[L_1, L_2] & \Gamma[L_2] \end{pmatrix}$$

is in general not singular, except when the following condition is fulfilled:

$$\begin{aligned} & \int_{t_1}^{t_1'} \frac{\ell \, dt}{\sqrt{\ell^2 - c^2 \cos^2 t}} \int_{t_2}^{t_2'} \frac{c \cos^2 t \, dt}{\sqrt{\ell^2 - c^2 \cos^2 t}} \\ & \qquad = \int_{t_2}^{t_2'} \frac{\ell \, dt}{\sqrt{\ell^2 - c^2 \cos^2 t}} \int_{t_1}^{t_1'} \frac{c \cos^2 t \, dt}{\sqrt{\ell^2 - c^2 \cos^2 t}}. \end{aligned} \tag{1}$$

Hence (see Chapter III, Section 3) if t_1, t_1', t_2, t_2' are such that condition (1) is not satisfied or only satisfied for (c, ℓ) in a set of probability zero, the pair (L_1, L_2) of arc lengths possesses an *absolutely continuous law*. For example, if $\cos t_1 \neq \cos t_2$, condition (1) is not satisfied when t_1' is in the neighborhood of t_1 and t_2' in the neighborhood of t_2.

2 Repeated samples: Discussion

Suppose we were to draw several ellipses using the preceding method. This example may serve to discuss the errors yielded by measurement devices in physics.

If we draw another ellipse with the same graduated rule and the same string, the two ellipses must be considered with correlated errors. It should be noted however that when the two ellipses are drawn with different instruments of the same type, the same conclusion is applicable. “Another” string may imply the following situations: the same strand cut at another length, a strand taken in another spool from the same storage, a strand taken in the same material. This observation shows that independence of errors tends to be the exceptional situation and will only be obtained by careful experimental procedures.

When errors are correlated, even should the quantity samples be independent in the sense of probability theory, they do not vanish by averaging. Henri Poincaré, mathematician and physicist, provided a mathematical model of this phenomenon of *error permanency* in his course on probability (Poincaré, 1912). The matter is to explain why:

> "with a meter divided into millimeters, as often as measures are repeated, a length will never be determined to within a millionth of a millimeter."[1]

This phenomenon is well-known by physicists, who have of course noted that over the entire history of experimental sciences, never has a quantity been precisely measured with rough instruments. Let us look at Poincaré's reasoning more closely.

First of all, he noted that the classical Gauss argument in favor of the normal law for the errors uses improper or excessive hypotheses. Let us recall Gauss' actual approach. Gauss considered that the quantity to be measured is random and can vary within the scope of the measurement device according to an *a priori* law. In modern language, let X be the random variable representing the quantity to be measured and μ be its law. The results of the measurement operations are other random variables $X_1, \ldots, X_n$; Gauss assumed that:

a) the conditional law of X_i given X be of the form

$$\mathbb{P}\{X_i \in E \mid X = x\} = \int_E \varphi(x_i - x)\, dx_i,$$

where φ is a smooth function;

b) the variables $X_1, \ldots, X_n$ be conditionally independent given X, in other words,

$$\mathbb{P}[X \in A, X_1 \in A_1, \ldots, X_n \in A_n] = \int_{x\in A} \int_{x_1 \in A_1} \cdots \int_{x_n \in A_n} \varphi(x_1 - x) \cdots \varphi(x_n - x)\, dx_1 \cdots dx_n\, d\mu(x).$$

He was then easily able to compute the conditional law of X given the results of measure $X_1, \ldots, X_n$:

$$\mathbb{P}[X \in A \mid X_1 = x_1, \ldots, X_n = x_n] = \int_{x\in A} \frac{\varphi(x_1 - x) \cdots \varphi(x_n - x)}{\int \varphi(x_1 - z) \cdots \varphi(x_n - z)\, d\mu(z)}\, d\mu(x),$$

which has a density with respect to μ:

$$\frac{\varphi(x_1 - x) \cdots \varphi(x_n - x)}{\int \varphi(x_1 - z) \cdots \varphi(x_n - z)\, d\mu(z)}.$$

Should this density be maximum at $x = \frac{1}{n}\sum_{i=1}^n x_i$ (which Gauss supposed as a starting point for his argument), one then obtains

$$\forall x_1 \cdots x_n \Big(x = \frac{1}{n} \sum_{i=1}^{n} x_i \Rightarrow \sum_{i=1}^{n} \frac{\varphi'(x_i - x)}{\varphi(x_i - x)} = 0 \Big).$$

[1] "Avec un mètre divisé en millimètres, on ne pourra jamais, si souvent qu'on répète les mesures, déterminer une longueur à un millionième de millimètre près." *op. cit.* p. 216.

This problem is purely analytical. Suppose $x, x_1, \ldots, x_n$ be scalar quantities for the sake of simplicity. We must have

$$\sum_i \frac{\partial}{\partial x_i}\left(\frac{\varphi'(x_i - x)}{\varphi(x_i - x)}\right) dx_i = 0 \quad \text{once} \sum_i dx_i = 0,$$

hence

$$\frac{\partial}{\partial x_1}\left(\frac{\varphi'(x_1 - x)}{\varphi(x_1 - x)}\right) = \cdots = \frac{\partial}{\partial x_n}\left(\frac{\varphi'(x_n - x)}{\varphi(x_n - x)}\right) = \text{constant}.$$

Thus

$$\frac{\varphi'(t - x)}{\varphi(t - x)} = a(t - x) + b.$$

Since φ is a probability density, Gauss obtained

$$\varphi(t - x) = \frac{1}{\sqrt{2\pi\sigma^2}} \exp\left(-\frac{(t - x)^2}{2\sigma^2}\right).$$

In order to explain the 'paradox' of non-vanishing errors by averaging, starting from this classical approach, Poincaré considered that given the true value x, the conditional law of X_i is not of the form

$$\varphi(y - x)\, dy,$$

but rather of the more general form

$$\varphi(y, x)\, dy.$$

He then wrote

$$\begin{aligned}&\mathbb{P}\big[X \in A, X_1 \in A_1, \ldots, X_n \in A_n\big]\\ &\quad = \int_{x \in A} \int_{x_1 \in A_1} \cdots \int_{x_n \in A_n} \varphi(x_1, x) \cdots \varphi(x_n, x)\, dx_1 \cdots dx_n\, d\mu(x),\end{aligned}$$

and then by setting

$$\int y\varphi(y, x)\, dy = x + \theta(x),$$

he remarked that the function θ was always constant in Gauss' approach and hence was zero in the absence of systematic error. However θ may satisfy $\int \theta(z)\, d\mu(z) = 0$, which expresses an absence of systematic error, without vanishing. In this case, under the conditional law $X = x$, which we denote $\mathbb{E}_x$,

$$\lim_{n \uparrow \infty} \mathbb{E}_x\left[\left(\frac{1}{n}\sum_{i=1}^n X_i - x - \theta(x)\right)^2\right] = 0$$

hence

$$\lim_{n\uparrow\infty} \mathbb{E}_x\left[\left(\frac{1}{n}\sum_{i=1}^{n} X_i - x\right)^2\right] = \theta^2(x)$$

due to the law of large numbers, with $X_1, \dots, X_n$ being independent under $\mathbb{E}_x$.

Averaging thus leaves an asymptotic error and the quantity cannot be known with this instrument to higher level of accuracy than $\theta^2(x)$.

It could be contested that Poincaré's explanation is based simply on the idea of a nonuniform parallax and that this could be overcome by careful calibration. Indeed, if we were to present the instrument with a perfectly-known quantity x_0, the average of a large number of measures would yield

$$\int y\varphi(y, x_0)\, dy = x_0 + \theta(x_0);$$

repeating this calibration for different values of x_0 will give the function $x \mapsto x+\theta(x)$ pointwise. It would be sufficient to inverse the function $I + \theta$ in order to obtain a measure to a desired level of precision, which contradicts Poincaré's principle of error permanency.

His example reveals having anticipated the potential objection: the function $I+\theta$ he obtains is a noninvertible step function. For a measure conducted with a graduated rule, he supposes that

$$\begin{array}{lll}
\text{if } n-\varepsilon < x \le n+\varepsilon & \text{then} & X_1 = n \\
\text{if } n+\varepsilon < x \le n+1-\varepsilon & \text{then} & X_1 = n \text{ with probability } \frac{1}{2} \\
 & \text{and} & X_1 = n+1 \text{ with probability } \frac{1}{2} \\
\text{if } n+1-\varepsilon < x \le n+1+\varepsilon & \text{then} & X_1 = n+1 \\
\vdots & &
\end{array}$$

and this conditional law gives for function θ and $I + \theta$ the following graphs.

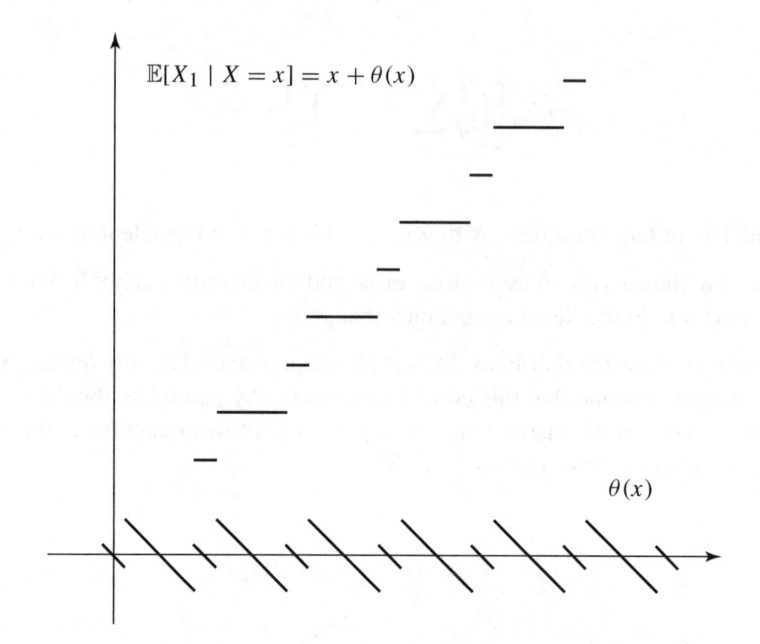

Let us emphasize one important point: in the Gauss model, the quantities that physically represent errors, i.e. random variables $X_1 - x, \dots, X_n - x$, are orthogonal under $\mathbb{E}_x$ since

$$\mathbb{E}_x\big[X_i\big] = x \quad \text{and} \quad \mathbb{E}_x\big[(X_1 - x)(X_2 - x)\big] = 0$$

by independence of the X_i's under $\mathbb{E}_x$. On the contrary, in Poincaré's model

$$\mathbb{E}_x\big[X_i\big] = x + \theta(x),$$

hence

$$\mathbb{E}_x\big[(X_1 - x)(X_2 - x)\big] = \mathbb{E}_x\big[X_1 - x\big]\mathbb{E}_x\big[X_2 - x\big] = \theta^2(x).$$

In other words, Poincaré suggested a model in which measures are independent but not errors. This layout is easily written using the language of error structures.

If the sample of a quantity is modeled by the probability space $(\mathbb{R}, \mathcal{B}(\mathbb{R}), m)$ and its error by the error structure $(\mathbb{R}, \mathcal{B}(\mathbb{R}), m, d, \gamma)$, a sequence of N independent samples can then be modeled on the probability space

$$(\mathbb{R}, \mathcal{B}(\mathbb{R}), m)^N.$$

To represent a sequence of repeated samples in which the errors are correlated, it suffices to consider an error structure

$$\big(\mathbb{R}^N, \mathcal{B}(\mathbb{R}^N), m^{\otimes N}, \mathbb{D}_N, \Gamma_N\big), \tag{2}$$

where Γ, restricted to one-argument functions coincides with γ, but contains rectangle terms.

When N varies, the error structures (2) constitute a projective system. Depending on the properties of this projective system (closability or non-closability) errors vanish or not by averaging. A detailed study is carried out in Bouleau [2001].

3 Calculation of lengths using the Cauchy–Favard method (exercise)

The arc length of a curve in the plane may be obtained by measuring the length of its projection on a straight line with director angle θ and then by averaging over θ. This result was proved by Cauchy for regular curves in 1832. He showed that for an arc parametrized by $t \in [0,1]$,

$$\int_0^1 \sqrt{\dot{x}^2(t)+\dot{y}^2(t)}\,dt = \frac{\pi}{2}\int_0^\pi\int_0^1 |\dot{x}(t)\cos\theta + \dot{y}(t)\sin\theta|\,dt\,\frac{d\theta}{\pi}. \tag{3}$$

A century later J. Favard extended this idea and proposed a measure for subsets of $\mathbb{R}^n$ (called Favard measure or integral-geometric measure) that coincides with the Hausdorff measure on rectifiable sets (Favard, 1932; Steinhaus, 1954; Federer, 1969).

To obtain the mean along different directions we may of course proceed by random samples.

Let us suppose we are interested in the length of a string modeled by

$$X_t = X_0 + \int_0^t \cos(\varphi + B_s)\,ds$$
$$Y_t = Y_0 + \int_0^t \sin(\varphi + B_s)\,ds,$$

where $(B_s)_{s\geq 0}$ is a standard Brownian motion and φ is a random variable uniformly distributed on the circle. As t varies from 0 to L, the length of arc $(X_t, Y_t)_{t\in[0,T]}$ is

$$\int_0^L \sqrt{\dot{X}^2(t)+\dot{Y}^2(t)}\,dt = L.$$

By means of very narrowly-drawn lines parallel to Oy, we can measure the total length of the projection of the string on Ox, say $P(\varphi,\omega)$:

$$P(\varphi,\omega) = \int_0^L \big|\cos(\varphi + B_s(\omega))\big|\,ds = \int_0^L |\dot{X}(t)|\,dt.$$

If this experiment were to be repeated many times independently, the expectation $\mathbb{E}P$ would be obtained by taking the average according to the law of large numbers. Now, from Fubini's Theorem since φ is uniformly distributed on the circle, we have

$$\begin{aligned}\mathbb{E}[P] &= \int_0^{2\pi} \mathbb{E}_\omega\left[\int_0^L \left|\cos(\varphi + B_s(\omega))\right| ds\right] \frac{d\varphi}{2\pi}\\ &= \int_0^L \int_0^{2\pi} |\cos(\varphi)| \frac{d\varphi}{2\pi}\, ds\\ &= \frac{2}{\pi} L.\end{aligned}$$

The length of the string is therefore yielded by the formula

$$L = \frac{\pi}{2}\mathbb{E}[P] \tag{4}$$

i.e. in denoting P_n the results of the independent samples,

$$L = \frac{\pi}{2} \lim_{N\uparrow\infty} \frac{1}{N} \sum_{n=1}^{N} P_n. \tag{5}$$

In order to model the errors, we must then choose hypotheses for the errors on both φ and $\left(B_s\right)_{s\geq 0}$ as well as hypotheses for the correlation of errors of different samples. The result for the asymptotic error

$$\lim_{N\uparrow\infty} \Gamma\left[\frac{1}{N}\sum_{n=1}^{N} P_n\right]$$

depends on the hypotheses for the correlation of errors, see Bouleau [2001]. We limit the discussion herein to remarks on the modeling of error on the string.

a) In the preceding model,

$$\begin{aligned}X(t) &= X_0 + \int_0^t \cos\left(\varphi + B_s\right) ds\\ Y(t) &= Y_0 + \int_0^t \sin\left(\varphi + B_s\right) ds, \quad 0 \leq t \leq L,\end{aligned}$$

the string is of class C^1 and t is its curvilinear abscissa. An Ornstein–Uhlenbeck type error on the Brownian motion then gives

$$\begin{aligned}P^\# &= -\int_0^L \operatorname{sign}\left(\cos(\varphi + B_s)\right) \sin\left(\varphi + B_s\right) \hat{B}_s\, ds\\ \Gamma[P] &= \int_0^L \int_0^L \operatorname{sign}\left(\cos(\varphi + B_s)\right) \sin\left(\varphi + B_s\right)\\ &\qquad \times \operatorname{sign}\left(\cos(\varphi + B_s)\right) \sin\left(\varphi + B_s\right) \cdot s \wedge t \cdot ds\, dt.\end{aligned}$$

In this model the string displays no resistance to bending since the direction of its tangent changes as rapidly as a Brownian motion.

b) In the following model,

$$X_t = X_0 + \int_0^t \cos(\varphi + V_s)\,ds$$

$$Y_t = Y_0 + \int_0^t \sin(\varphi + V_s)\,ds, \quad 0 \le t \le L \le 1,$$

with

$$V_s = \int_0^1 u \wedge s\,dB_s$$

the string resists bending and is of class C^2. An Ornstein–Uhlenbeck type error corresponds to a modification in the curvature by means of adding a small perturbation.

In this model, we can consider more violent errors, e.g. with an error operator of the type

$$\Gamma\left[\int_0^1 f(s)\,dB_s\right] = \int_0^1 f'^2(s)\,ds$$

(see Chapter V, Section 2.4) we then obtain

$$\Gamma[V_s, V_t] = s \wedge t$$

and the conditional law of the error on some characteristic of the string given the tangent process (process V_s, $s \in [0,1]$) is the same as in case a).

4 Temperature equilibrium of a homogeneous solid (exercise)

Let U be a bounded open set in $\mathbb{R}^3$ with boundary ∂U representing a homogeneous body. Let A_1 and A_2 be two disjoint subsets of ∂U whose temperature is permanently maintained at θ_1 [resp. θ_2], with the remainder of the boundary remaining at a temperature θ_0.

Suppose θ_1, θ_2 are erroneous random variables, whose image error structure is

$$\left([a_1, b_1] \times [a_2, b_2], \mathcal{B}, \frac{dx_1}{b_1 - a_1} \times \frac{dx_2}{b_2 - a_2}, \mathbb{D}, \Gamma\right)$$

with

$$\Gamma[u](x_1, x_2) = u_1'^2 + u_2'^2.$$

We will now study the error on temperature $\theta(M)$ at a point M of the body.

Let f be the function defined on ∂U by

$$f = \theta_0 + (\theta_1 - \theta_0)1_{A_1} + (\theta_2 - \theta_0)1_{A_2}.$$

If we suppose that A_1 and A_2 are Borel sets, by the Perron–Wiener–Brelot theorem (see Brelot, 1959 and 1997), the bounded Borel function f is resolutive; thus if $H[f]$ is the generalized solution to the Dirichlet problem associated with f, we have

$$\theta(M) = H[f](M) = \theta_0 + (\theta_1 - \theta_0)H\big[1_{A_1}\big](M) + (\theta_2 - \theta_0)H\big[1_{A_2}\big](M).$$

Let us once again introduce the sharp operator defined on $\mathbb{D}$ by

$$u^{\#} = \big(u_1'\tilde{\theta}_1,\ u_2'\tilde{\theta}_2\big)$$

with

$$\tilde{\theta}_i = \frac{\hat{\theta}_i - \hat{\mathbb{E}}\,\hat{\theta}_i}{\hat{\mathbb{E}}\big(\hat{\theta}_i - \hat{E}\,\hat{\theta}_i\big)^2}, \qquad i = 1, 2.$$

The linearity of operator H yields

$$\big(H[f](M)\big)^{\#} = H\big[f^{\#}\big](M).$$

If M_1 and M_2 are two points in U, the matrix

$$\underline{\underline{\Gamma}} = \begin{pmatrix} \Gamma\big[\theta(M_1)\big] & \Gamma\big[\theta(M_1), \theta(M_2)\big] \\ \Gamma\big[\theta(M_1), \theta(M_2)\big] & \Gamma\big[\theta(M_2)\big] \end{pmatrix}$$

is singular if and only if the vectors

$$\big(\theta(M_1)\big)^{\#} \quad \text{and} \quad \big(\theta(M_2)\big)^{\#}$$

are proportional.

Since $\big(\theta(M)\big)^{\#} = \big(H\big[1_{A_1}\big](M)\tilde{\theta}_1,\ H\big[1_{A_2}\big](M)\tilde{\theta}_2\big)$ the matrix $\underline{\underline{\Gamma}}$ is singular if and only if

$$\begin{vmatrix} H\big[1_{A_1}\big](M_1) & H\big[1_{A_1}\big](M_2) \\ H\big[1_{A_2}\big](M_1) & H\big[1_{A_2}\big](M_2) \end{vmatrix} = 0. \tag{6}$$

Let us fix point M_1.

- If M_2 does not belong to the surface defined in (6) the temperatures in M_1 and M_2 have uncorrelated errors; by the image energy density property (see Chapter III, Proposition III.16), the law of the pair $\big(\theta(M_1), \theta(M_2)\big)$ is absolutely continuous with respect to the Lebesgue measure on $\mathbb{R}^2$.
- If M_2 belongs to the surface (6), the random variables

$$\begin{aligned} \theta(M_1) &= \theta_0 + (\theta_1 - \theta_0)H\big[1_{A_1}\big](M_1) + (\theta_2 - \theta_0)H\big[1_{A_2}\big](M_1) \\ \theta(M_2) &= \theta_0 + (\theta_1 - \theta_0)H\big[1_{A_2}\big](M_2) + (\theta_2 - \theta_0)H\big[1_{A_2}\big](M_2) \end{aligned}$$

  are linked by a deterministic linear relation and the law of the pair $\big(\theta(M_1), \theta(M_2)\big)$ is carried by a straight line.

More generally, if the temperature at the boundary is of the form

$$f(M) = a_1 f_1(M) + \cdots + a_k f_k(M), \quad M \in \partial U,$$

where the functions $f_1, \ldots, f_k$ are integrable for the harmonic measure μ_{M_0} at a point $M_0 \in U$ (hence for the harmonic measure of every point in U) and linearly independent in $L^1(\mu_{M_0})$ and if the error on f is given by the fact that the a_i's are random and erroneous with uncorrelated errors, then the temperatures at k points $\theta(M_1), \ldots, \theta(M_k)$ have uncorrelated errors and an absolutely continuous joint law on $\mathbb{R}^k$, except if points $M_1, \ldots, M_k$ satisfy the condition that the vectors

$$\big(H[f_1](M_i), \ldots, H[f_k](M_i)\big), \quad i = 1, \ldots, k,$$

are linearly dependent, in which case the law of the k-uple $\theta(M_1), \ldots, \theta(M_k)$ is carried by a hyperplane in $\mathbb{R}^k$.

In order to handle the case where the boundary function f is given by an infinite dimensional expansion, let us simply consider a body modeled by the unit disc in $\mathbb{R}^2$.

A) Suppose f is given by

$$f(\varphi) = a_0 + \sum_{n\geq 1} a_n \sqrt{2} \cos n\varphi + \sum_{m\geq 1} b_m \sqrt{2} \sin m\varphi$$

on the complete orthonormal system $(1, \sqrt{2}\cos n\varphi \sqrt{2}\sin m\varphi)$ of $L^2(d\varphi)$ and assume that a_i and b_j are independent erroneous random variables defined on $(\Omega, \mathfrak{A}, \mathbb{P}, \mathbb{D}, \Gamma)$ such that $\mathbb{P}$-a.s. only a finite number of the a_n's and b_m's do not vanish (as in Chapter V, Section 4 and Chapter VII, Section 4.2), with

$$\begin{aligned} \Gamma\big[a_j\big] = a_i^2, \quad \Gamma\big[b_j\big] = b_j^2 \quad \forall i, j \\ \Gamma\big[a_i, a_j\big] = \Gamma\big[b_i, b_j\big] = 0 \quad \forall i \neq j \\ \Gamma\big[a_i, b_k\big] = 0 \quad \forall i, k. \end{aligned}$$

Let a point of the open disc be represented by its complex affix

$$z = \rho(\cos\alpha + i\sin\alpha).$$

Then the solution of the Dirichlet problem is explicit:

$$H[f](z) = a_0 + \sum_{n\geq 1} a_n \sqrt{2} \rho^n \cos n\alpha + \sum_{m\geq 1} b_m \sqrt{2} \rho^m \sin m\alpha$$

hence

$$\big(H[f](z)\big)^{\#} = a_0^{\#} + \sum_{n\geq 1} a_n^{\#} \sqrt{2} \rho^n \cos n\alpha + \sum_{m\geq 1} b_m^{\#} \sqrt{2} \rho^m \sin m\alpha.$$

From this formula the error on any function of the temperature in the disc may be easily calculated.

For example, let C be an arc in the disc defined in polar coordinates. The integral

$$I = \int_\alpha^\beta H[f]\big(\rho(t)e^{it}\big)\sqrt{\rho'^2(t)+\rho^2(t)}\,dt \Big/ \int_\alpha^\beta \sqrt{\rho'^2(t)+\rho^2(t)}\,dt$$

represents the average of the temperature on C. We have

$$\Gamma[I] = \frac{1}{L(C)^2}\bigg\{L(C)^2 a_0^2 + \sum_{n\geq 1} 2a_n^2\left(\int_\alpha^\beta \rho^n(t)\cos nt\sqrt{\rho'^2+\rho^2}\,dt\right)^2 + \sum_{m\geq 1} 2b_m^2\left(\int_\alpha^\beta \rho^m(t)\sin mt\sqrt{\rho'^2+\rho^2}\,dt\right)^2\bigg\}$$

where $L(C) = \int_\alpha^\beta \sqrt{\rho'^2+\rho^2}\,dt$ denotes the length of the arc.

B) Now suppose the boundary function f is given by

$$f(\varphi) = g_0 + \sum_{n\geq 1} g_n \int_0^\varphi \sqrt{2}\cos n\psi \frac{d\psi}{2\pi} + \sum_{m\geq 1} \tilde{g}_m \int_0^\varphi \sqrt{2}\sin m\psi \frac{d\psi}{2\pi},$$

where the $(g_0; g_n, n \geq 1; \tilde{g}_m, m \geq 1)$ are independent, reduced Gaussian variables. By the classical construction of Brownian motion (see Chapter VI, Section 2.1), we observe that $f(\varphi)$ is a *Brownian bridge*, on which like for Brownian motion, several error structures are available.

The temperature at $z = \rho(\cos\alpha + i\sin\alpha)$ is

$$H[f](z) = g_0 + \sum_{n\geq 1} g_n \int_0^\alpha \sqrt{2}\rho^n(\cos n\beta)\frac{d\beta}{2\pi} + \sum_{m\geq 1} \tilde{g}_m \int_0^\alpha \sqrt{2}\rho^m(\sin m\beta)\frac{d\beta}{2\pi}.$$

This expression may be seen as a stochastic integral with respect to the Brownian bridge f.

If we set

$$h(\varphi) = 1 + \sum_{n\geq 1}\left(\int_0^\alpha \sqrt{2}\rho^n(\cos n\beta)\frac{d\beta}{2\pi}\right)\sqrt{2}\cos n\varphi + \sum_{m\geq 1}\left(\int_0^\alpha \sqrt{2}\rho^m(\sin m\beta)\frac{d\beta}{2\pi}\right)\sqrt{2}\sin m\varphi$$

we obtain

$$H[f](z) = \int h\,df.$$

Hence, if we choose an Ornstein–Uhlenbeck type error on f, i.e. such that

$$\Gamma\left[\int h\,df\right] = \int h^2(\varphi)\,d\varphi,$$

we derive

$$\begin{aligned}\Gamma\big[H[f](z)\big] &= \int_0^{2\pi} h^2(\varphi)\,d\varphi \\ &= 1 + \sum_{n\geq 1}\left(\int_0^\alpha \sqrt{2}\rho^n(\cos n\beta)\frac{d\beta}{2\pi}\right)^2 + \sum_{m\geq 1}\left(\int_0^\alpha \sqrt{2}\rho^m(\sin m\beta)\frac{d\beta}{2\pi}\right)^2.\end{aligned}$$

Similar calculations can be performed with other error structures on the Brownian bridge, e.g. with

$$\Gamma\left[\int h\,df\right] = \int h'^2(\varphi)\,d\varphi$$

since $z \to H[f](z)$ is a $\mathcal{C}^\infty$-function in the disc, it remains in the domain of Γ and we obtain

$$\begin{aligned}\Gamma\big[H[f](z)\big] = &\sum_{n\geq 1}\left(\int_0^\alpha \sqrt{2}\rho^n n(\sin n\beta)\frac{d\beta}{2\pi}\right)^2 \\ &+ \sum_{m\geq 1}\left(\int_0^\alpha \sqrt{2}\rho^m m(\cos m\beta)\frac{d\beta}{2\pi}\right)^2.\end{aligned}$$

In this problem, the quantities studied were linearly dependent on the data. Such is no longer the case in the next example.

5 Nonlinear oscillator subject to thermal interaction: The Grüneisen parameter

In this section, we demonstrate that the effect of thermal agitation on a small macroscopic body produces inaccuracies, which may be represented thanks to the language of error structures.

We first consider the case of a harmonic oscillator. The position and velocity are erroneous and the error structure is obtained from the Boltzmann–Gibbs law using linearity of the system. This result can also be found by means of a finer modeling along the lines of the historical work of Ornstein and Uhlenbeck using the Langevin equation.

The case of a nonlinear oscillator may be approached similarly by assuming that the oscillator has a slow proper movement in absence of the thermal interaction. The result can also be obtained herein by finer modeling based on a stochastic differential equation.

Finally we relate the *bias* provided by error calculus in the nonlinear case to both the *thermal expansion* of crystals and the Grüneisen parameter.

5.1 Harmonic oscillator. We now consider a one-dimensional oscillator governed by the equation

$$m\ddot{X}_t + rX_t = F(t), \tag{7}$$

where m is the inertial coefficient, rX_t the return force and $F(t)$ the applied force.

If the applied force is zero and if the oscillator is subject only to the thermal interaction, the Boltzmann–Gibbs law states that the position and velocity display according to the following probability law:

$$\nu(dx, dv) = C \exp\left\{-\frac{\mathcal{H}}{kT}\right\} dx\, dv$$

where $\mathcal{H} = \frac{1}{2}m\dot{x}^2 + \frac{1}{2}rx^2$ is the Hamiltonian of the oscillator, k the Boltzmann constant and T the absolute temperature. Put otherwise

$$\nu(dx, dv) = N\left(0, \frac{kT}{r}\right)(dx) \times N\left(0, \frac{kT}{m}\right)(dv).$$

In the general case where $F(t)$ does not vanish, thanks to system linearity, we may write

$$X_t = y_t + x_t,$$

where y_t would be the "cold" movement due to the applied force and x_t the thermal movement.

If we consider x_t as an error, i.e. using the notation of Chapter I, and if we set

$$\begin{aligned} \Delta X_t &= x_t \\ \Delta \dot{X}_t &= \dot{x}_t, \end{aligned}$$

we obtain the variances of the errors

$$\begin{cases} \operatorname{var}\big[\Delta X_t \mid X_t = x, \dot{X}_t = v\big] \# \mathbb{E}\big[x_t^2\big] = \dfrac{kT}{r} \\ \operatorname{var}\big[\Delta \dot{X}_t \mid X_t = x, \dot{X}_t = v\big] \# \mathbb{E}\big[\dot{x}_t^2\big] = \dfrac{kT}{m} \\ \operatorname{covar}\big[\Delta X_t, \Delta \dot{X}_t \mid X_t = x, \dot{X}_t = v\big] \# \operatorname{covar}\big[x_t, \dot{x}_t\big] = 0 \end{cases} \tag{8}$$

and for the biases

$$\begin{cases} \mathbb{E}\big[\Delta X_t \mid X_t = x, \dot{X}_t = v\big] \# \mathbb{E}\big[x_t\big] = 0 \\ \mathbb{E}\big[\Delta \dot{X}_t \mid X_t = x, \dot{X}_t = v\big] \# \mathbb{E}\big[\dot{x}_t\big] = 0. \end{cases} \tag{9}$$

In other words, we are led to setting

$$
(10)\qquad \begin{cases} \Gamma[X_t] = \dfrac{kT}{r} \\ \Gamma[\dot{X}_t] = \dfrac{kT}{m} \\ \Gamma[X_t, \dot{X}_t] = 0 \\ A[X_t] = 0 \\ A[\dot{X}_t] = 0. \end{cases}
$$

As mentioned several times previously, knowledge of the three objects (a priori probability measure, Γ and A) is excessive in determining an error structure since when the probability measure is fixed, operators Γ and A can be defined by each other. Actually system (10) is compatible here with the closable pre-structure

$$
\left( ]a, b[, \mathcal{B}(]a, b[), \frac{dx}{b-a}, \mathcal{C}_K^\infty(]a, b[), u \to \frac{kT}{r} u'^2 \right)
$$
$$
\times \left( ]c, d[, \mathcal{B}(]c, d[), \frac{dy}{d-c}, \mathcal{C}_K^\infty(]c, d[), v \to \frac{kT}{m} v'^2 \right)
$$

with $a < b$, $c < d$ in $\mathbb{R}$. The bounds a, b and c, d in general have no particular physical meaning and we may consider the structure

$$
\left(\mathbb{R}^2, \mathcal{B}(\mathbb{R}^2), dx\, dy, H_0^1(\mathbb{R}^2), \Gamma\right)
$$

with

$$
\Gamma[w](x, y) = \frac{kT}{r}\left(\frac{\partial w}{\partial x}\right)^2 + \frac{kT}{m}\left(\frac{\partial m}{\partial y}\right)^2,
$$

which is a Dirichlet structure (see Bouleau–Hirsch [1991]) with a σ-finite measure as an a priori measure instead of a probability measure, this difference however would not fundamentally change the reasoning.

5.2 The Ornstein–Uhlenbeck approach with the Langevin equation. In a famous article, Ornstein and Uhlenbeck [1930] (see also Uhlenbeck–Wang [1945]) rediscover the Boltzmann–Gibbs law thanks to a complete analysis of the random process describing how the oscillator moves. The authors start from the so-called Langevin equation, first proposed by Smoluchowski.

The idea herein is to represent the movement of the free oscillator (without applied forces) in thermal interaction by the equation

$$
(11)\qquad m\ddot{x} + f\dot{x} + rx = U(t).
$$

In this equation, m and r are the coefficients of equation (7), $U(t)$ is a white noise, i.e. the time derivative, with a suitable sense, of a mathematical Brownian motion with variance $\sigma^2 t$, and the coefficient f adds a friction term to equation (7). This new term is compulsory if we want the Markov process of the pair (position, velocity) to possess an invariant measure, which is a probability measure in accordance with the Boltzmann–Gibbs law. The term $f\dot{x}$ must be considered as a friction due to the thermal interaction. This viscosity effect may be understood by the fact that the oscillator receives more impacts towards the left-hand side when moving to the right and vice-versa.

The study of equation (11) is classical and can be conducted within the framework of stationary processes theory or Markov processes theory, (see Bouleau [2000], Chapter 5, ex. 5.11 and Chapter 7, §7.4.3). If the intensity of white noise is σ^2, which can be written $U(t) = \sigma \frac{dB_t}{dt}$, equation (11) has a stationary solution x_t, which is a centered Gaussian process with spectral measure

$$\frac{\sigma^2}{\left(\frac{r}{m} - \lambda^2\right)^2 + \frac{f^2}{m^2}\lambda^2}\,\frac{d\lambda}{2\pi}. \tag{12}$$

thus, in the weakly-damped case, its covariance is

$$\begin{cases} K_X(t) = \mathbb{E}\big[x_{t+s}\,x_s\big] \\ \quad = \dfrac{\sigma^2}{4\xi\omega_0^3} e^{-\xi\omega_0|t|}\left(\cos\omega_1|t| + \dfrac{\xi\omega_0}{\omega_1}\sin\omega_1|t|\right) \end{cases} \tag{13}$$

with $\omega_0^2 = \frac{r}{m}$, $\xi = \frac{f}{2m\omega_0}$ and $\omega_1 = \omega_0\sqrt{1-\xi^2}$. The overdamped case is obtained by setting $\omega_1 = i\omega'$ and the aperiodic case is obtained when $\omega_1 \to 0$. We shall consider here, for example, the weakly damped case.

From its spectral measure (13), we deduce that the stationary Gaussian process x_t is of class $\mathcal{C}^1$ in $L^2(\mathbb{P})$ (where $\mathbb{P}$ is the probability measure serving to define the process).

The study within the Markovian framework may be conducted by setting $\frac{dx_t}{dt} = v_t$. The Langevin equation then becomes

$$\begin{cases} dx_t = v_t\,dt \\ dv_t = \sigma\,dB_t - \left(\dfrac{f}{m}v_t + \dfrac{r}{m}\,x_t\right)dt \\ x_0 = a_0, \quad v_0 = b_0. \end{cases} \tag{14}$$

This linear equation defines a Gaussian Markov process with generator

$$\mathcal{A}[w](x,v) = \frac{1}{2}\sigma^2\frac{\partial^2 w}{\partial v^2} - \left(\frac{f}{m}v + \frac{r}{m}x\right)\frac{\partial w}{\partial v} + v\frac{\partial w}{\partial x}.$$

Although the diffusion matrix $\begin{pmatrix} 0 & 0 \\ 0 & \sigma \end{pmatrix}$ is singular, the Kolmogoroff–Hörmander condition is satisfied and the pair (x_t, v_t) has a density $\forall t > 0$. It can be shown (see

Ornstein–Uhlenbeck [1930] and Wang–Uhlenbeck [1945]) that as $t \to \infty$, the pair (x_t, v_t) converges in distribution to the measure

$$(15) \qquad \nu(dx, dv) = N\left(0, \frac{\sigma^2}{4\xi\omega_0^3}\right)(dx) \times N\left(0, \frac{\sigma^2}{4\xi\omega_0}\right)(dv).$$

Even though (x_t) and (v_t) are bound together by the deterministic relation

$$x_{t_1} = x_{t_0} + \int_{t_0}^{t_1} v_s \, ds,$$

asymptotically for large t, x_t and v_t are independent. This phenomenon is typical in statistical thermodynamics.

For equation (15) to be compatible with the Boltzmann–Gibbs law

$$(16) \qquad \nu(dx, dv) = C \exp\left\{-\frac{1}{2}\frac{r}{kT}x^2 - \frac{1}{2}\frac{m}{kT}v^2\right\} dx\, dv,$$

the white noise constant σ must be linked with the viscosity coefficient f by the relation

$$(17) \qquad \sigma = \frac{\sqrt{2fkT}}{m}.$$

Remarks. 1) The proportionality relation (17) between σ^2 and f allows for the energy brought to the system by thermal impacts to be evacuated by the viscosity such that an invariant probability measure appears. In the case of the Kappler experiment of a small mirror hung from a torsion thread in a gas, σ^2 and f linked by (17) depend on the pressure of this gas.

2) Let us mention that if we had started with a damped oscillator, instead of an undamped oscillator, governed by the equation

$$\mu\ddot{X} + \varphi\dot{X} + \rho X = F(t),$$

(for instance, in the case of an RLC electric circuit), we would have to face a dissipative system, which does not obey the Hamiltonian mechanics (except by introducing an ad hoc dissipative function, see Landau–Lipschitz [1967]). For such a system, the Boltzmann–Gibbs law, which introduces the Hamiltonian, is not directly applicable. Knowledge of the evolution of such a system at temperature T in a given environment must describe the modifications due to thermal energy being returned by the system. Nevertheless, if we assume that the result may be modeled by an equation of the type

$$m_1\ddot{X} + f_1\dot{X} + r_1 X = U(t),$$

an asymptotic probability measure of the same form as (15) will appear with different coefficients, i.e. the Boltzmann–Gibbs measure of an equivalent asymptotic Hamiltonian system. The new proportionality relation (17) between f_1 and σ^2 expresses a general relation between correlation and damping, known as fluctuation-dissipation theorems (see Kubo et al. [1998]). ◇

Let us now return to the undamped oscillator subject to thermal agitation and satisfying, as a consequence, both equation (14) and relation (17). If we take $\nu(dx, dv)$ as the initial measure, the process (x_t, v_t) is Markovian, Gaussian and stationary.

When the oscillator moves macroscopically the thermal interaction permanently modifies the position and velocity. The mathematical description of these reciprocal influences is precisely the probability law of a stationary Gaussian process like x_t.

Representing the thermal interaction by an error structure. From the magnitude of the Boltzmann constant, we may consider the thermal perturbation as an error. For this purpose we introduce an error structure on process x_t along with the following notation: let $\Omega = \mathcal{C}^1(\mathbb{R})$, $\mathfrak{A} = \mathcal{B}(\mathcal{C}^1(\mathbb{R}))$ and let $\mathbb{P}$ be the probability measure that makes coordinate maps x_t a centered Gaussian process with covariance (13). In other words, $\mathbb{P}$ is the law of (x_t). Since x_t is the solution of a linear equation with respect to the Gaussian white noise, the perturbation

$$x_.(\omega) \longrightarrow \sqrt{e^{-\varepsilon}}x_.(\omega) + \sqrt{1-e^{-\varepsilon}}\,\hat{x}_.(\hat{\omega}),$$

where $(\hat{x}_t(\omega))$ is a copy of (x_t) defined on $(\hat{\Omega}, \hat{\mathfrak{A}}, \hat{\mathbb{P}})$, defines a semi-group P_ε by

$$P_\varepsilon[G](\omega) = \hat{\mathbb{E}}\left[G\big(\sqrt{e^{-\varepsilon}}x_.(\omega) + \sqrt{1-e^{-\varepsilon}}\hat{x}_.(\hat{\omega})\big)\right]. \tag{18}$$

The reasoning expressed in Chapter II and Chapter VI, Section 2 applies herein and shows that P_ε is a strongly-continuous contraction semigroup on $L^2(\Omega, \mathfrak{A}, \mathbb{P})$, symmetric with respect to $\mathbb{P}$. We define in this manner an error structure $(\Omega, \mathfrak{A}, \mathbb{P}, \mathbb{D}, \Gamma)$, satisfying

$$\left\{\begin{aligned}
\Gamma[x_t] &= \mathbb{E}[x_t^2] = \frac{kT}{r}\\
\Gamma[x_t, x_s] &= \mathbb{E}[x_t x_s] = \frac{kT}{r}e^{-\xi\omega_0|t-s|}\Big(\cos\omega_1|t-s| + \frac{\xi}{\sqrt{1-\xi^2}}\sin\omega_1|t-s|\Big)\\
\Gamma[\dot{x}_t] &= \mathbb{E}[\dot{x}_t^2] = \frac{kT}{m}\\
\Gamma[\dot{x}_t, \dot{x}_s] &= \mathbb{E}[\dot{x}_t\dot{x}_s] = \frac{kT}{m}e^{-\xi\omega_0|t-s|}\Big(\cos\omega_1|t-s| - \frac{\xi}{\sqrt{1-\xi^2}}\sin\omega_1|t-s|\Big)\\
\Gamma[x_t, \dot{x}_s] &= \mathbb{E}[x_t\dot{x}_s] = \frac{kT}{m\omega_1}e^{-\xi\omega_0|t-s|}\sin\omega_1|t-s|.
\end{aligned}\right. \tag{19}$$

This structure is preserved by time translation. Its properties may be summarized by assigning it the following sharp:

$$(x_t)^{\#} = \hat{x}_t,$$

where $\hat{x}_t$ is a copy of x_t.

Let us now suppose that in addition to the thermal interaction, we act upon the system by means of the applied force $F(t)$. The movement

$$m\ddot{X}_t + f\dot{X}_t + rX_t = U(t) + F(t)$$

may be decomposed into $X_t = y_t + x_t$ and supposing y_t is not erroneous, we therefore derive

$$(20)\qquad \begin{cases} \Gamma[X_t] = \dfrac{kT}{r} \\ \Gamma[X_t, X_s] = \dfrac{kT}{r} e^{-\xi\omega_0|t-s|}\left(\cos\omega_1|t-s| + \dfrac{\xi}{\sqrt{1-\xi^2}}\sin\omega_1|t-s|\right) \\ \Gamma[\dot{X}_t] = \dfrac{kT}{m} \\ \Gamma[\dot{X}_t, \dot{X}_s] = \dfrac{r}{m}\Gamma[X_t, X_s] \\ \Gamma[X_t, \dot{X}_s] = \dfrac{kT}{m\omega_1} e^{-\xi\omega_0|t-s|}\sin\omega_1|t-s|. \end{cases}$$

Remarks. 1) Let us note that the approximation from considering the thermal interaction to be an error implies that

$$\Gamma[Z] = \mathrm{var}[Z]$$

if Z is in $\mathbb{D}$ and a linear function of the process $(x_t)_{t\in\mathbb{R}}$, but no longer so when Z is a more general random variable. For instance, the quadratic function of kinetic energy $\frac{1}{2}m\dot{X}_t^2$ has an error

$$\Gamma\left[\frac{1}{2}m\dot{X}_t^2\right] = \frac{m^2}{4}\dot{X}_t^2\Gamma[\dot{X}_t] = \frac{1}{2}m\dot{X}_t^2\frac{kT}{2},$$

which is not constant but proportional to the kinetic energy.

2) If we consider diffusion (14) with the invariant measure as initial law, i.e.

$$\begin{cases} x_t = x_0 + \displaystyle\int_0^t v_s\,ds \\ v_t = v_0 + \sigma B_t - \displaystyle\int_0^t \left(\frac{f}{m}v_s + \frac{r}{m}x_s\right) ds \end{cases}$$

where (x_0, v_0) is a random variable of law $\nu(dx, dv)$ independent of $\left(B_t\right)_{t\geq 0}$, we observe that introducing an Ornstein–Uhlenbeck structure on B_t and an independent error structure on (x_0, v_0) such that

$$\Gamma[x_0] = \frac{kT}{r}, \quad \Gamma[v_0] = \frac{kT}{m}, \quad \Gamma[x_0, v_0] = 0,$$

which can be summarized by the sharp

$$\begin{aligned}\left(B_t\right)^\# &= \hat{B}_t \\ x_0^\# &= \hat{x}_0 \\ v_0^\# &= \hat{v}_0\end{aligned}$$

where, as usual, $\hat{B}_t$, $\hat{x}_0$, $\hat{v}_0$ are copies of B_t, x_0, v_0, give for x_t exactly the above structure $(\Omega, \mathfrak{A}, \mathbb{P}, \mathbb{D}, \Gamma)$.

5.3 The nonlinear oscillator. We now consider a one-dimensional nonlinear oscillator governed by the equation

$$m\ddot{X} + r(X) = F(t). \tag{21}$$

The device involved may be a small mirror, as in the Kappler experiment, with a nonlinear return force (Figure 1), an elastic pendulum with two threads (Figure 2) or circular (Figure 3), or a vibrating molecule in a solid (Figure 4).

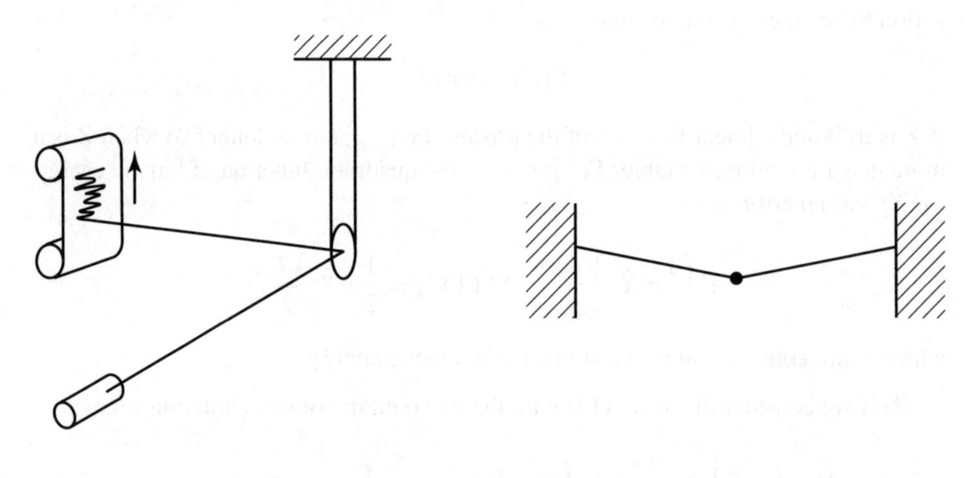

Figure 1 Figure 2

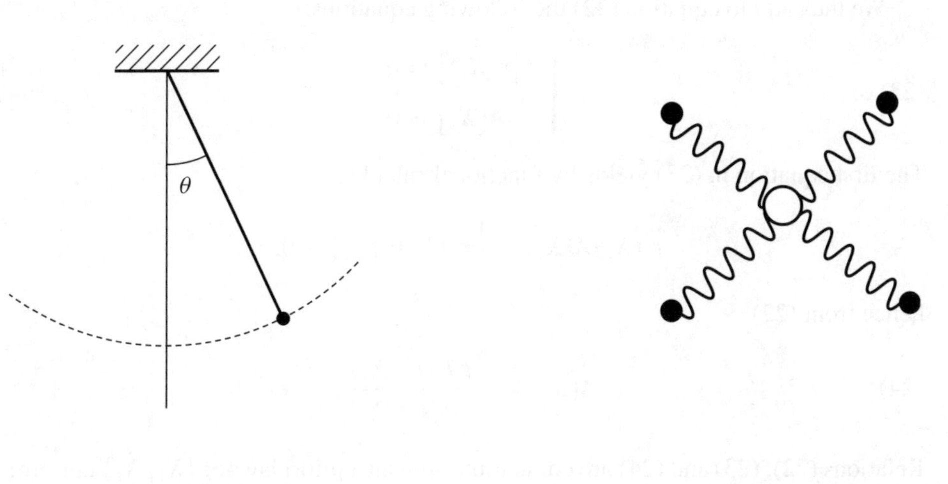

Figure 3 Figure 4

We can no longer decompose X_t into the form $X_t = y_t + x_t$ as before due to nonlinearity.

We shall nevertheless consider that the proper movement under the applied force $F(t)$ without thermal interaction is slow with respect to the time of transition to thermal equilibrium. We are seeking the error structure to be applied on X and $\dot{X}$ in order to provide a correct account of the error due to thermal interaction.

This hypothesis allows us to linearize equation (21) for equilibrium density and we may then write the Boltzmann–Gibbs principle by stating that the thermal movement follows the law

$$\nu(dx, dv) = C \exp\left\{-\frac{1}{2}\frac{r'(X_t)(x - X_t)^2}{kT} - \frac{1}{2}\frac{m(v - \dot{X}_t)^2}{kT}\right\}.$$

We are thus led to set

$$(22) \qquad \begin{cases} \Gamma[X_t] = \dfrac{kT}{r'(X_t)} \\ \Gamma[\dot{X}_t] = \dfrac{kT}{m} \\ \Gamma[X_t, \dot{X}_t] = 0. \end{cases}$$

The function $x \to r(x)$ is assumed to be regular and strictly increasing.

Regarding the biases, it is only natural to consider both that the error on $\dot{X}_t$ is centered and that the return force

$$r(X_t)$$

is well-described macroscopically by the function r itself, i.e. that $r(X_t)$ has no bias: $A\big[r(X_t)\big] = 0$.

We thus add to equation (22) the following equations:

$$\text{(23)} \qquad \begin{cases} A\big[r(X_t)\big] = 0 \\ A\big[\dot{X}_t\big] = 0. \end{cases}$$

The first equation in (23) yields by functional calculus

$$r'(X_t)A[X_t] + \frac{1}{2}r''(X_t)\Gamma[X_t] = 0,$$

hence from (22)

$$\text{(24)} \qquad A[X_t] = -\frac{kT}{2}\frac{r''(X_t)}{r'^2(X_t)}.$$

Relations (22), (23) and (24) are compatible with an a priori law for $\big(X_t, \dot{X}_t\big)$ uniform on a rectangle and give rise to the product error structure defined as the closure of the following pre-structure for $Z_t = \big(X_t, \dot{X}_t\big)$

$$\text{(25)} \qquad \begin{cases} \left(]a,b[\times]c,d[,\ \mathcal{B}(]a,b[\times]c,d[),\ \frac{dx}{b-a}\frac{dy}{d-c},\ \mathcal{C}_K^\infty(]a,b[\times]c,d[),\ \Gamma\right) \\ \Gamma[w](x,y) = \frac{kT}{r'(x)}\left(\frac{\partial w}{\partial x}\right)^2 + \frac{kT}{m}\left(\frac{\partial w}{\partial y}\right)^2 \end{cases}$$

whose bias operator is

$$\text{(26)} \qquad A[w](x,y) = \frac{1}{2}\frac{kT}{r'(x)}\frac{\partial^2 w}{\partial x^2} - \frac{kT}{2}\frac{r''(x)}{r'^2(x)}\frac{\partial w}{\partial x} + \frac{1}{2}\frac{kT}{m}\frac{\partial^2 w}{\partial y^2}.$$

The same comment as that following relation (10) may be forwarded here concerning the bounds a, b and c, d which in general have no precise physical meaning.

As usual, the random variable in equations (25) and (26) is represented by the identity map from $\mathbb{R}^2$ into $\mathbb{R}^2$, i.e. the error structure (25) is the image error structure of Z_t.

a) Example from Figure 2:

$$r(x) = 2\lambda x - \frac{2a\lambda x}{\sqrt{a^2+x^2}},$$

where λ is the elastic constant of the threads and $2a$ is the distance of the supports,

$$r'(x) = 2\lambda\left[1 - \frac{a}{\sqrt{a^2+x^2}} + \frac{ax^2}{(a^2+x^2)^{3/2}}\right] > 0$$

$$r''(x) = \frac{6\lambda a^3 x}{(a^2+x^2)^{5/2}}$$

such that

$$A[X_t] = -\frac{kT}{2}\frac{6\lambda a^3 X_t}{(a^2+X_t^2)^{5/2}}\cdot\frac{1}{r'^2(X_t)}.$$

$A[X]$ and X have opposite signs. If the object is submitted to its weight, the equilibrium position is slightly above what it would take without thermal interaction.

b) Example of the single pendulum (Figure 3): The governing equation is

$$m\ell\ddot{\theta} + mg\sin\theta = F \quad \theta \in \left]-\frac{\pi}{2}, \frac{\pi}{2}\right[$$

thus

$$\Gamma[\theta] = \frac{kT}{mg\cos\theta}$$

$$A[\theta] = \frac{kT}{2}\frac{\sin\theta}{mg\cos^2\theta}.$$

The bias and θ have same signs.

c) Cubic oscillator:

$$r(x) = \omega^2 x(1+\beta x^2), \quad \beta > 0$$

$$\Gamma[X] = \frac{kT}{\omega^2}\cdot\frac{1}{(1+3\beta X^2)}$$

$$A[X] = -\frac{kT}{2}\frac{6\beta X}{\omega^2(1+3\beta X^2)^2}.$$

The bias and X have opposite signs.

d) Nonsymmetric quadratic oscillator:

$$r(x) = \alpha x - \beta x^2 \quad \text{if } x < \frac{\alpha}{2\beta}$$

$$r(x) = \frac{\alpha}{2\beta} \quad \text{if } x \geq \frac{\alpha}{2\beta}.$$

In other words,

$$r(x) = \int_0^x (\alpha - 2\beta y)^+ \, dy.$$

Supposing for the sake of simplicity that x varies in $]a, b[\subset \left[-\infty, \frac{\alpha}{2\beta}\right[$,

$$\Gamma[X] = \frac{kT}{\alpha - 2\beta X}$$

$$A[X] = \frac{kT}{2}\frac{2\beta}{(\alpha - 2\beta X)^2}$$

the bias is positive.

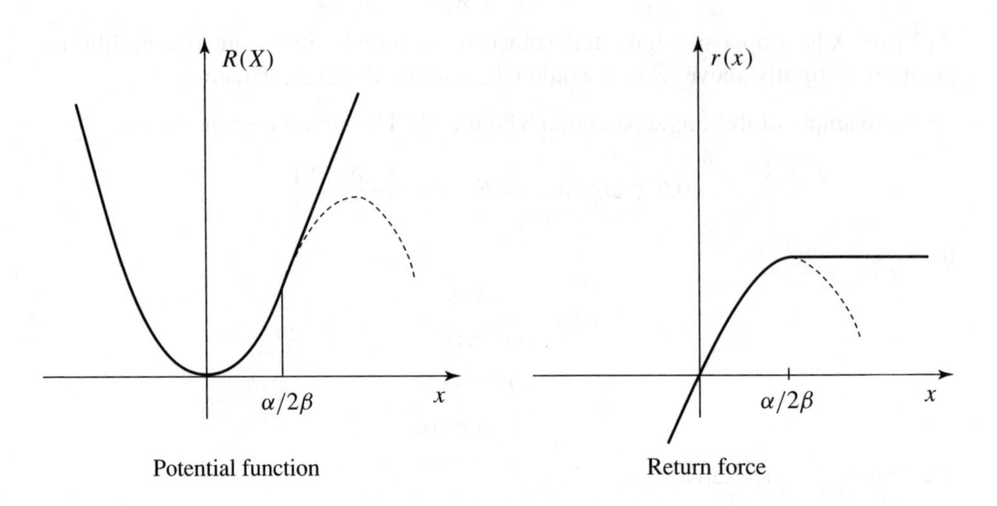

Potential function

Return force

e) Oscillator with constant bias: If $r(x) = \alpha + \beta \log x$, $\alpha \in \mathbb{R}$, $\beta > 0$ we have

$$\Gamma[X] = \frac{kT}{\beta} X$$

$$A[X] = \frac{kT}{2\beta}.$$

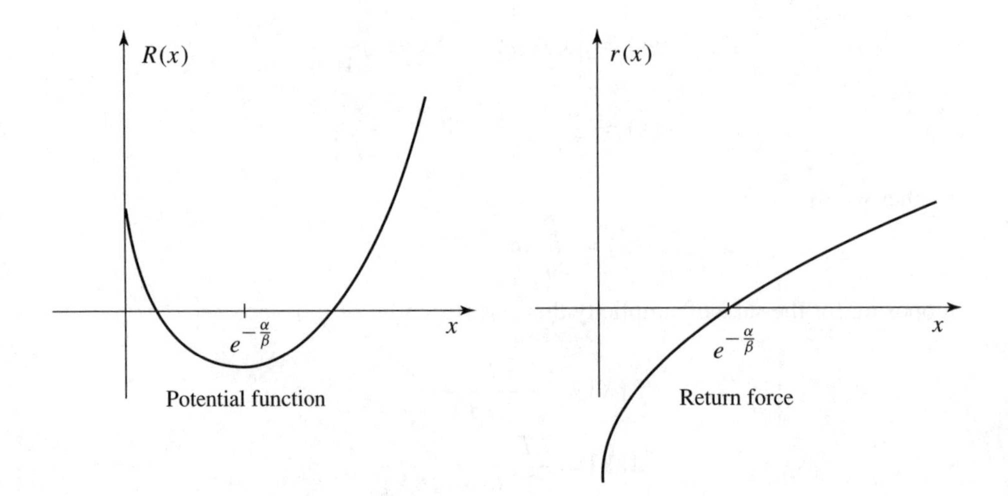

Potential function

Return force

5.4 Ornstein–Uhlenbeck approach in the nonlinear case. The system in (14) is changed into

$$
(27) \qquad \begin{cases} dX_t = \dot{X}_t \, dt \\ d\dot{X}_t = \sigma \, dB_t - \left(\frac{f}{m} \dot{X}_t + \frac{r(X_t)}{m} \right) dt \\ X_0 = a_0, \quad \dot{X}_0 = b_0 \end{cases}
$$

which is an Itô equation. Using the generator of this diffusion, it is easy to verify (as soon as f and σ are linked by relation (17)) that the probability measure

$$
\nu(dx, dv) = C e^{-\frac{R(x)}{kT} - \frac{1}{2} \frac{m}{kT} v^2} \, dx \, dv,
$$

where $R(x) = \int_0^x r(y) \, dy$ is the potential associated with the return force $r(x)$, is invariant by the semigroup of the diffusion (27).

If (a_0, b_0) follows law ν, the process $(X_t, \dot{X}_t)$ is stationary. Let $\Omega = \mathcal{C}^1(\mathbb{R}_+)$, $\mathfrak{A} = \mathcal{B}(\mathbb{R}_+)$ and let $\mathbb{P}$ be the probability measure on $(\Omega, \mathfrak{A})$, such that the coordinate mappings build a realisation of the above process $(X_t, \dot{X}_t)$.

In order to describe the thermal interaction as an error, we introduce an error on $(\Omega, \mathfrak{A}, \mathbb{P})$ in such a way that

a) on (a_0, b_0) which follows law ν, we consider an error governed by the sharp operator $(a_0^\#, b_0^\#)$ independently of that of the Brownian motion,

b) on the Brownian motion, we consider the same error as in the linear case, i.e.

$$
(B_t)^\# = \hat{B}_t.
$$

Then from equation

$$
(28) \qquad \begin{cases} X_t = X_0 + \int_0^t \dot{X}_s \, ds \\ \dot{X}_t = \dot{X}_0 + \int_0^t \sigma \, dB_s - \int_0^t \left(\frac{f}{m} \dot{X}_s + \frac{r(X_s)}{m} \right) ds \end{cases}
$$

the following relations can be derived:

$$
(29) \qquad \begin{cases} X_t^\# = X_0^\# + \int_0^t \dot{X}_s^\# \, ds \\ \dot{X}_t^\# = \dot{X}_0^\# + \sigma \hat{B}_t - \int_0^t \left(\frac{f}{m} \dot{X}_s^\# + \frac{r'(X_s) X_s^\#}{m} \right) ds. \end{cases}
$$

This system is linear in $(X_t^\#, \dot{X}_t^\#)$ and easily solved matricially:

Let

$$Z_t = \begin{pmatrix} X_t \\ \dot{X}_t \end{pmatrix}, \quad \underline{\underline{\sigma}} = \begin{pmatrix} 0 & 0 \\ 0 & \sigma \end{pmatrix}, \quad D(X_s) = \begin{pmatrix} 0 & 1 \\ -\frac{r'(X_s)}{m} & -\frac{f}{m} \end{pmatrix}$$

and $W_s = \begin{pmatrix} 0 \\ \hat{B}_s \end{pmatrix}$. System (29) may then be written as follows:

$$Z_t^{\#} = Z_0^{\#} + \int_0^t \underline{\underline{\sigma}}\, dW_s + \int_0^t D(X_s) Z_s^{\#}\, ds. \tag{30}$$

The matrices $M_t = \left(m_t^{ij}\right)$ and $N_t = \left(n_t^{ij}\right)$, solutions to

$$M_t = I + \int_0^t D(X_s) M_s\, ds$$

$$N_t = I - \int_0^t N_s D(X_s)\, ds,$$

satisfy $M_t N_t = N_t M_t = I$, hence $N_t = M_t^{-1}$, and the solution to equation (30) is

$$Z_t^{\#} = M_t \left(Z_0^{\#} + \int_0^t M_s^{-1} \underline{\underline{\sigma}}\, dW_s \right). \tag{31}$$

The error matrix on Z_t follows

$$\underline{\underline{\Gamma}}[Z_t] = \begin{pmatrix} \Gamma[X_t] & \Gamma[X_t, \dot{X}_t] \\ \Gamma[X_t, \dot{X}_t] & \Gamma[\dot{X}_t] \end{pmatrix}$$

$$= M_t \underline{\underline{\Gamma}}[Z_0] \tilde{M}_t + \sigma^2 M_t \int_0^t \begin{pmatrix} (n_s^{12})^2 & n_s^{12} n_s^{22} \\ n_s^{12} n_s^{22} & (n_s^{22})^2 \end{pmatrix} ds\, \tilde{M}_t \tag{32}$$

where $\tilde{M}_t$ denotes the transposed matrix of M_t.

The absolute continuity criterion (Proposition III.16) yields the following result (see Bouleau–Hirsch, Chapter IV, §2.3).

Proposition. *For r Lipschitz and $\alpha \neq 0$, the solution $\left(X_t, \dot{X}_t\right)$ of* (28) *possesses an absolute continuous law on $\mathbb{R}^2$ for $t > 0$.*

Proof. The fact that $\det\left[\underline{\underline{\Gamma}}[Z_t]\right] > 0$ $\mathbb{P}$-a.s. is easily derived from the relation (32) by taking into account both the continuity of $s \to N_s$ and the fact that $n_0^{22} = 1$, $n_0^{12} = 0$ and $n_s^{12} = -\frac{m_s^{12} n_s^{22}}{m_s^{11}}$ does not vanish for sufficiently small $s \neq 0$. ◇

The method may be extended to the case where the oscillator is subject to an applied force $F(t)$:

$$m\ddot{X}_t + r\big(X_t\big) = F(t). \tag{33}$$

The method consists of solving (33) knowing $F(t)$ and initial conditions X_0, $\dot{X}_0$, and then determining the matrices M_t and N_t by

$$M_t = I + \int_0^t D(X_s)M_s\,ds$$
$$N_t = I - \int_0^t N_s D(X_s)\,ds,$$

where

$$D(X_s) = \begin{pmatrix} 0 & 1 \\ -\frac{r'(X_s)}{m} & -\frac{f}{m} \end{pmatrix}.$$

The operator Γ is then given by

$$\begin{pmatrix} \Gamma[X_t] & \Gamma[X_t,\dot{X}_t] \\ \Gamma[X_t,\dot{X}_t] & \Gamma[\dot{X}_t] \end{pmatrix} = M_t \begin{pmatrix} \Gamma[X_0] & \Gamma[X_0,\dot{X}_0] \\ \Gamma[X_0,\dot{X}_0] & \Gamma[\dot{X}_0] \end{pmatrix} + M_t \int_0^t N_s \begin{pmatrix} 0 & 0 \\ 0 & \sigma^2 \end{pmatrix} \tilde{N}_s\,ds\,\tilde{M}_t.$$

This equation generally requires an approximate numerical resolution. An analytical solution is available, for instance, when $F(t)$ produces the following forced oscillation:

$$X(t) = X_0 + \frac{\dot{X}_0}{\omega}\sin\omega t$$

i.e. when

$$\begin{aligned} F(t) &= m\ddot{X}_t + f\dot{X}_t + r(X_t) \\ &= -m\dot{X}_0\omega\sin\omega t + f\dot{X}_0\cos\omega t + r\left(X_0 + \frac{\dot{X}_0}{\omega}\sin\omega t\right). \end{aligned}$$

The coefficients of matrices M_t and N_t are then defined in terms of Mathieu functions.

In the *quasi-static case* where the movement due to $F(t)$ is slow, we may further simplify and consider that X_t maintains the same value whilst the thermal interaction installs the invariant measure, i.e. to consider $D(X_t)$ constant in equation (30). The solution to (30) is thus that found in the linear case with new parameters dependent on X_t; hence for fixed ω, the process

$$Z_t^{\#}(\omega,\hat{\omega}) = \left(X_t^{\#}(\omega,\hat{\omega}), \dot{X}_t^{\#}(\omega,\hat{\omega})\right)$$

follows the law

$$N\left(0, \frac{\sigma^2}{4\tilde{\xi}\,\tilde{\omega}_0^3}\right)(dx) \times N\left(0, \frac{\sigma^2}{4\tilde{\xi}\,\tilde{\omega}_0}\right)(dv),$$

where $\tilde{\xi} = \frac{f}{2m\tilde{\omega}_0}$ and $\tilde{\omega}_0^2 = \frac{r'(X_t)}{m}$. From these hypotheses, we once again derive (22):

$$(34) \qquad \begin{cases} \Gamma[X_t] = \dfrac{kT}{r'(X_t)}, \quad \Gamma[\dot{X}_t] = \dfrac{kT}{m} \\ \Gamma[X_t, \dot{X}_t] = 0. \end{cases}$$

5.5 Thermal expansion of crystals and the Grüneisen parameter. Let us now consider a crystal of linear dimension L and volume V. If left unconstrained, the volume of this body will change with temperature (it generally expands with increasing T, but not always).

It is clear that if the crystal were considered as a lattice of harmonic oscillators (in neglecting entropic effects), an increment of temperature does not affect the crystal volume. In order to produce the common experiment whereby the dilatation coefficient is generally positive, the model must necessarily involve the nonlinearity of elementary oscillators (see Tabor [1991]).

Let us now return to our model of the nonlinear oscillator (see Sections 5.3 and 5.4 above)

$$m\ddot{X} + r(X) = F$$

and let us represent the thermal interaction by an error structure as explained above. In the quasi-static case, the thermal movement follows a linear dynamics governed by the equation

$$m\ddot{x} + f\dot{x} + r'(X)x = U(t)$$

such that $\frac{r'(X)}{m} = \omega^2$ is the square of the thermal movement pulsation. This term varies with X, and the quantity γ defined by

$$\frac{d\omega}{\omega} = \gamma\frac{dV}{V} = 3\gamma\frac{dX}{X}$$

is dimensionless. It links together the proportional variation of pulsation and the proportional variation of volume (which is three times that of the linear dimension) and is called the *Grüneisen parameter*.

With such a definition, the Grüneisen parameter would be negative for typical bodies and, in keeping with most authors, we shall change the signs and set

(35)
$$\gamma = -\frac{d\ln\omega}{d\ln V} = -\frac{1}{3}\frac{d\ln\omega}{d\ln X} = -\frac{1}{6}\frac{d\ln\omega^2}{d\ln X}.$$

This parameter is a measurable characteristic of the nonlinearity of the oscillator and can be related to thermodynamic quantities. Under specific hypotheses it relates to the volume dilatation coefficient α_{th} by

$$\alpha_{th} = \gamma\frac{C_v}{3K},$$

where K is the bulk modulus and C_v the heat capacity per volume unit.

We thus obtain

$$\gamma = -\frac{1}{6}\frac{r''(X)}{r'(X)}X.$$

According to our hypothesis that the return force $r(X)$ is without bias $A[r(X)] = 0$, which yields

$$r'(X)A[X] + \frac{1}{2}r''(X)\Gamma[X] = 0,$$

we note that the Grüneisen parameter is related to the error structure of the thermal interaction by the relation

(36)
$$\gamma = \frac{1}{3}\frac{A[X]/X}{\Gamma[X]/X^2}.$$

Up to a factor of $\frac{1}{3}$, this parameter is therefore the ratio of proportional bias of the error to proportional variance of the error. Since we know that bias $A[X]$ and variance $\Gamma[X]$ have the same order of magnitude proportional to temperature (see Sections 5.1 and 5.2), we observe that, in an initial approximation, γ does not depend on temperature.

Oscillators with constant γ. For ordinary materials under usual conditions, γ lies between 1 and 3.

We can select the function $r(x)$ such that γ is constant. By setting

$$r(x) = \alpha + \frac{\beta}{x^{6\gamma-1}}, \quad \alpha, \beta > 0;\ \gamma > \frac{1}{3},$$

we immediately obtain that $\gamma = -\frac{1}{6}\frac{r''(x)}{r'(x)}x$ is indeed the Grüneisen parameter.

Such a model corresponds to a potential with the following shape:

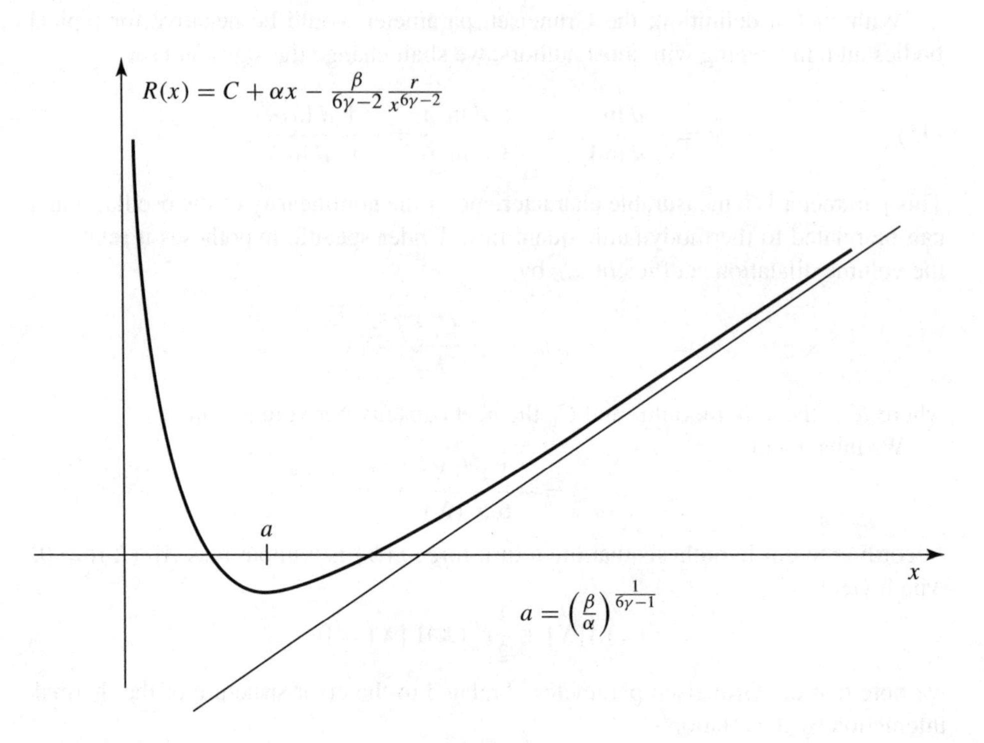

Oscillators with proportional bias. The condition that $\frac{A[X]}{X}$ is constant leads to two types of models:

i) $r(x) = \alpha - \dfrac{\beta}{x}, \ \alpha, \beta > 0,$

$$\Gamma[X] = \frac{kT}{\beta}X^2 \qquad A[X] = kT\frac{X}{\beta}.$$

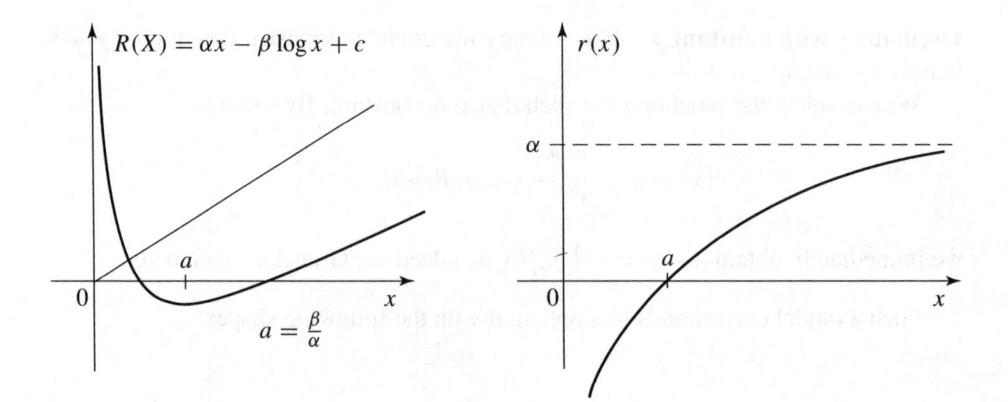

ii) $r(x) = \alpha \operatorname{Arctan} \beta x + C$

$$\alpha > 0, \ \beta > 0, \quad c \in \left]-\frac{\pi}{2}, 0\right[$$

whose potential is

$$R(x) = \alpha x \operatorname{Arctan} \beta x + \frac{\alpha}{\beta} \log \cos \operatorname{Arctan} \beta x + Cx + C_1.$$

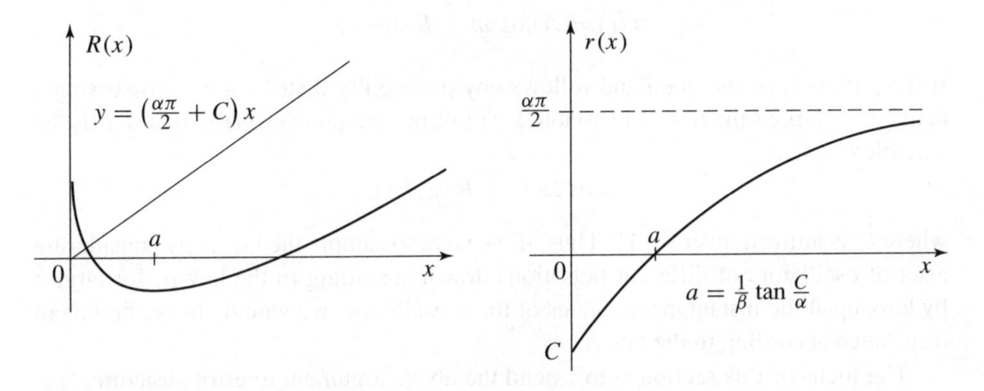

When the molecular oscillators of a crystal correspond to these models, the proportional biases do not depend on X, hence the thermal dilatation coefficient is preserved whenever a tensile force is applied to the crystal.

Comments

We stayed in this example concerning oscillators and thermal expansion of crystals, inside classical mechanics. It seems that similar calculations may be performed in changing thermal incertitude into quantic incertitude. For example in the case of a weakly quantic anharmonic oscillator, the shape of the obtained potential might be a slight change of the potential of the particle looked as classical.

Calculations of fluctuations in classical or quantic statistical mechanics seem to be also a field of possible application of error calculus based on Dirichlet forms. This is particularly apparent in the chapter concerning fluctuations (Chapter XII, p. 414–478) of the famous book of Landau and Lifchitz [1967], in which the square of the mean fluctuational displacement of a quantity X is denoted $\overline{(\Delta X)^2}$ and handled according to the functional calculus of $\Gamma[X]$.

6 Natural error structures on dynamic systems

In order to justify or explain that some systems or devices are governed by simple, generally uniform, probability laws (such that the entire number between 1 and 36 obtained by the roulette wheel, or the little planets repartition on the Zodiac), Henri Poincaré proposed an original argument, which has since been renamed "the arbitrary functions method". This argument consists of showing that for a large class of probability laws on the initial conditions and the parameters, the system will always over the long run move toward the same probability distribution.

The simplest example is that of the harmonic oscillator. Let us consider a simple pendulum with small oscillations or an oscillating electric circuit without damping, governed by the equation

$$x(t) = A\cos\omega t + B\sin\omega t.$$

If the pulsation is uncertain and follows any probability distribution μ possessing a density, for large t the random variable $x(t)$ follows the same probability law ρ as the variable:

$$A\cos 2\pi U + B\sin 2\pi U,$$

where U is uniform over $[0, 1]$. Thus, if we were to sample the law μ by considering a set of oscillators of different pulsations drawn according to the law μ, for large t by looking at the instantaneous states of these oscillators we would always find them distributed according to the law ρ.

The focus of this section is to extend the above argument to error structures, i.e. to show that for a large class of error structures on the data, some dynamic systems are asymptotically governed by simple error structures.

We will first introduce a notion of convergence that extends the narrow convergence (or convergence in distribution) to error structures.

Definition VIII.1. Let E be a finite-dimensional differential manifold and let $S_\lambda = (E, \mathcal{B}(E), \mathbb{P}_\lambda, \mathbb{D}_\lambda, \Gamma_\lambda)$, $\lambda \in \Lambda$, and $S = (E, \mathcal{B}(E), \mathbb{P}, \mathbb{D}, \Gamma)$ be error structures such that $C^1 \cap \operatorname{Lip}(E) \subset \mathbb{D}_\lambda$ $\forall\lambda$ and $C^1 \cap \operatorname{Lip}(E) \subset \mathbb{D}$. S_λ is said to *converge D-narrowly* (Dirichlet-narrowly) to the error structure S as $\lambda \to \lambda_0$, if $\mathbb{P}_\lambda$ tends to $\mathbb{P}$ narrowly,

$$\forall u \in \mathcal{C}_b(E), \quad \mathbb{E}_\lambda[u] \to \mathbb{E}[u],$$

and if the Dirichlet forms converge on $C^1 \cap \operatorname{Lip}$-functions:

$$\forall v \in \mathcal{C}^1 \cap \operatorname{Lip}(E)$$

$$\mathbb{E}_\lambda\big[\Gamma_\lambda[v]\big] \to \mathbb{E}\big[\Gamma[v]\big].$$

Example. Let $(\Omega_0, \mathfrak{A}_0, \mathbb{P}_0, \mathbb{D}_0, \Gamma_0)$ be an error structure and let X_0 be a centered random variable in $\mathbb{D}_0^d$. Let $X_1, X_2, \dots, X_n, \dots$ be a sequence of copies of X_0 defined on the product structure:

$$\Sigma = \big(\Omega_0, \mathfrak{A}_0, \mathbb{P}_0, \mathbb{D}_0, \Gamma_0\big)^{\mathbb{N}^*}.$$

Let S_n be the image structures of Σ on $\mathbb{R}^d$ by the variables

$$V_n = \frac{1}{\sqrt{n}}(X_1 + \cdots + X_n)$$

then an extended form of the central limit theorem (Bouleau–Hirsch, Chapter V, Thm. 4.2.3) states that structures S_n converge D-narrowly to the structure S closure of

$$\left(\mathbb{R}^d, \mathcal{B}(\mathbb{R}^d), N(0, M), \mathcal{C}^1 \cap \mathrm{Lip}, \Gamma\right),$$

where $N(0, M)$ is the centered normal law of dispersion matrix $M = \left(\mathbb{E}\left[X_{0_i} X_{0_j}\right]\right)_{ij}$ and where Γ is given by $\forall u \in \mathcal{C}^1 \cap \mathrm{Lip}(\mathbb{R}^d)$

$$\Gamma[u] = \sum_{i,j=1}^{d} \mathbb{E}\left[\Gamma_0\left[X_{0_i}, X_{0_j}\right]\right] \frac{\partial u}{\partial x_i} \frac{\partial u}{\partial x_j}.$$

The error structure S has a normal law and a Γ operator with constant coefficients, meaning it is an Ornstein–Uhlenbeck structure. This finding explains the importance of such a structure in the applications. ◇

6.1 Poincaré–Hopf style limit theorems. We begin with theorems in dimension one. Let $S = (\Omega, \mathfrak{A}, \mathbb{P}, \mathbb{D}, \Gamma)$ be an error structure and let X be in $\mathbb{D}$.

Let us consider the random variable

$$X_t = tX \pmod 1,$$

that we consider with values in the torus $\mathbb{T}_1 = \mathbb{R}/\mathbb{Z}$ equipped with its circle topology. Let us denote the image error structure by X_t as

$$S_t = \left(\mathbb{T}_1, \mathcal{B}(\mathbb{T}_1), \mathbb{P}_{X_t}, \mathbb{D}_{X_t}, \Gamma_{X_t}\right)$$

which we then renormalize into

$$\tilde{S}_t = \left(\mathbb{T}_1, \mathcal{B}(\mathbb{T}_1), \mathbb{P}_{X_t}, \mathbb{D}_{X_t}, \frac{1}{t^2}\Gamma_{X_t}\right).$$

We then have

Proposition VIII.2. *If the law of X possesses a density, the structure $\tilde{S}_t$ converges D-narrowly as $t \to \infty$ toward the structure*

$$\left(\mathbb{T}_1, \mathcal{B}(\mathbb{T}_1), \lambda_1, H^1(\mathbb{T}_1), u \to u'^2 \mathbb{E}\Gamma[X]\right)$$

where λ_1 is the Haar measure on $\mathbb{T}_1$.

Proof. a) The part of the result concerning narrow convergence of the probability laws is classical (see Poincaré [1912], Hopf [1934], Engel [1992]) and based on the following lemma.

Lemma VIII.3 *Let μ_t be a family of probability measures and μ a probability measure on the torus $\mathbb{T}_1$. Then $\mu_t \to \mu$ narrowly if and only if*

$$\hat{\mu}_t(k) \to \hat{\mu}(k) \quad \forall k \in \mathbb{Z},$$

where

$$\hat{\mu}_t(k) = \int_{\mathbb{T}_1} e^{2i\pi kx}\, d\mu_t(x),\ \hat{\mu}(k) = \int_{\mathbb{T}_1} e^{2i\pi kn}\, d\mu(x).$$

Proof of the lemma. The necessity of the condition stems immediately from the fact that functions $x \to e^{2i\pi kx}$ are bounded and continuous. The sufficiency is due to the fact that any continuous function from $\mathbb{T}_1$ in $\mathbb{R}$ may be uniformly approximated by linear combinations of these functions using Weierstrass' theorem. ◇

Let us return to the proposition. We have

$$\mathbb{E}e^{2i\pi kX_t} = \mathbb{E}e^{2i\pi ktX} = \varphi_X(2\pi kt)$$

and by the Riemann–Lebesgue lemma $\lim_{|u|\uparrow\infty} \varphi_X(u) = 0$, hence

$$\lim_{t\to\infty} \mathbb{E}e^{2i\pi kX_t} = 0 \quad \forall k \neq 0.$$

From the lemma, this implies that X_t converges in distribution to the uniform law on the torus (i.e. Haar's measure).

b) Let $F \in \mathcal{C}^1 \cap \mathrm{Lip}(\mathbb{T}_1)$,

$$\Gamma_{X_t}[F](y) = \mathbb{E}\big[\Gamma[F(X_t)] \mid Y_t = y\big]$$

and

$$\mathbb{E}_{X_t}\left[\frac{1}{t^2}\Gamma_{X_t}[F]\right] = \frac{1}{t^2}\mathbb{E}\big[\Gamma[F(X_t)]\big].$$

Considering F to be a periodic function with period 1 from $\mathbb{R}$ into $\mathbb{R}$ and using functional calculus yields

$$\frac{1}{t^2}\mathbb{E}\big[\Gamma[F(X_t)]\big] = \mathbb{E}\big[F'^2(X_t)\Gamma[X]\big].$$

If $\Gamma[X] = 0$ $\mathbb{P}$-a.s., the right-hand side is zero and the proposition is proved. If $\mathbb{E}\Gamma[X] \neq 0$, since $\Gamma[X] \in L^1(\mathbb{P})$ and $F' \in \mathcal{C}_b$, it suffices to apply part a) with the probability measure $\frac{1}{\mathbb{E}\Gamma[X]}\Gamma[X] \cdot \mathbb{P}$ in order to obtain the result. ◇

Since we know that functional calculus remains valid in dimension 1 for Lipschitz functions, the proof shows that this convergence still holds for Lipschitz functions.

The map $X \to tX \pmod 1$ actually also erases the starting point and the correlations with the initial situation, as shown by the following:

Proposition VIII.4. *Let $(\Omega, \mathfrak{A}, \mathbb{P}, \mathbb{D}, \Gamma)$ be an error structure and X, Y, Z be random variables in $\mathbb{D}$, with X possessing a density. We set*

$$X_t = tX + Y \pmod 1,$$

considered with values in $\mathbb{T}_1$.

Let S_t be the image error structure by (X_t, X, Y, Z)

$$S_t = \left(\mathbb{T}_1 \times \mathbb{R}^3, \mathcal{B}, \mathbb{P}_{X_t,X,Y,Z}, \mathbb{D}_{X_t,X,Y,Z}, \Gamma_{X_t,X,Y,Z}\right)$$

and let $\tilde{S}_t$ be its renormalization

$$\tilde{S}_t = \left(\mathbb{T}_1 \times \mathbb{R}^3, \mathcal{B}, \mathbb{P}_{X_t,X,Y,Z}, \mathbb{D}_{X_t,X,Y,Z}, \frac{1}{t^2}\Gamma_{X_t,X,Y,Z}\right).$$

$\tilde{S}_t$ then converges D-narrowly to the product structure

$$\left(\mathbb{T}_1, \mathcal{B}(\mathbb{T}_1), \lambda_1, H^1(\mathbb{T}_1), u \to u'^2\mathbb{E}\Gamma[X]\right) \times \left(\mathbb{R}^3, \mathcal{B}(\mathbb{R}^3), \mathbb{P}_{X,Y,Z}, L^2, 0\right).$$

Proof. a) Regarding the convergence of probability measures, we have:

Lemma VIII.5. *Let μ_t and μ be probability measures on $\mathbb{T}_1 \times \mathbb{R}^3$. For $k \in \mathbb{Z}$ and $(u_1, u_2, u_3) \in \mathbb{R}^3$ we set*

$$\hat{\mu}_t(k, u_1, u_2, u_3) = \int e^{2i\pi ks + iu_1x_1 + iu_2x_2 + iu_3x_3}\, d\mu_t(s, x_1, x_2, x_3).$$

and we similarly define $\hat{\mu}$ from μ. Then $\mu_t \to \mu$ narrowly if and only if $\forall k, u_1, u_2, u_3$

$$\hat{\mu}_t(k, u_1, u_2, u_3) \to \hat{\mu}(k, u_1, u_2, u_3).$$

This result is general for the Fourier transform on locally compact groups (see Berg–Forst [1975], Chapter I, Thm. 3.14). ◇

As for the proposition, let $k, u_1, u_2, u_3 \in \mathbb{Z} \times \mathbb{R}^3$ and let us consider the expression

$$A_t = \mathbb{E}\left[e^{2i\pi k(tX+Y) + iu_1X + iu_2Y + iu_3Z}\right].$$

By setting $\xi = 2\pi kY + u_1X + u_2Y + u_3Z$, this expression can be written as follows:

$$A_t = \mathbb{E}\Big[e^{2i\pi ktX}(\cos\xi)\big(1_{\{\cos\xi>0\}} + 1_{\{\cos\xi=0\}} + 1_{\{\cos\xi<0\}}\big)\Big] \\ + i\mathbb{E}\Big[e^{2i\pi ktX}(\sin\xi)\big(1_{\{\sin\xi>0\}} + 1_{\{\sin\xi=0\}} + 1_{\{\sin\xi<0\}}\big)\Big].$$

Since X has a density under $\mathbb{P}$, it also has a density under the probability

$$\frac{(\cos\xi)1_{\{\cos\xi>0\}}}{\mathbb{E}\big[(\cos\xi)1_{\{\cos\xi>0\}}\big]}\cdot\mathbb{P}.$$

Hence, according to the Riemann–Lebesgue lemma, $A_t \to 0$ as $t\to\infty$ if $k\neq 0$. From Lemma VIII.5, the quadruplet (X_t, X, Y, Z) converges in distribution to $\lambda_1\times\mathbb{P}_{X,Y,Z}$.

b) Let $F\in\mathcal{C}^1\cap\mathrm{Lip}(\mathbb{T}_1\times\mathbb{R}^3)$. We have

$$\Gamma\big[F(X_t,X,Y,Z)\big] = F_1'^2\Gamma[X_t] + F_2'^2\Gamma[X] + F_3'^2\Gamma[Y] + F_4'^2\Gamma[Z] \\ + 2F_1'F_2'\Gamma[X_t,X] + 2F_1'F_3'\Gamma[X_t,X] + 2F_1'F_4\Gamma[X_t,Z] \\ + 2F_2'F_3'\Gamma[X,Y] + 2F_2'F_4'\Gamma[X,Z] + 2F_3'F_4'\Gamma[Y,Z].$$

By dominated convergence we observe that $\mathbb{E}\big[\frac{1}{t^2}\Gamma[F(X_t,X,Y,Z)]\big]$ has the same limit when $t\to\infty$ as $\mathbb{E}\big[F_1'^2(X_t,X,Y,Z)\Gamma[X]\big]$. But if $\mathbb{E}\Gamma[X]\neq 0$, under the law $\frac{\Gamma[X]}{\mathbb{E}\Gamma[X]}\cdot\mathbb{P}$, the variable X still possesses a density and by application of a)

$$\mathbb{E}\left[\frac{1}{t^2}\Gamma[F(X_t,X,Y,Z)]\right] \to \mathbb{E}\big[\Gamma[X]\big]\int F_1'^2(\alpha,x,y,z)\,d\lambda_1(\alpha)\,d\mathbb{P}_{X,Y,Z}(x,y,z)$$

which proves the proposition. ◇

6.2 Multidimensional cases, examples. The Riemann–Lebesgue lemma is a general property of characteristic functions of probability measures on locally compact Abelian groups possessing a density with respect to the Haar measure, (see Berg–Forst [1975], Chapter 1). This lemma allows extending the preceding results to more general cases. For instance, we have the following:

Proposition VIII.6. *Let* $(\Omega,\mathfrak{A},\mathbb{P},\mathbb{D},\Gamma)$ *be an error structure, X, Y variables in* $\mathbb{D}^d$ *and Z a variable in* $\mathbb{D}^q$*, with X possessing a density. We set $X_t = tX+Y$ with values in the d-dimensional torus* $\mathbb{T}_d$*. The renormalized image error structure by* (X_t,X,Y,Z),

$$\left(\tilde{S}_t = \mathbb{T}_d\times\mathbb{R}^{2d+q},\mathcal{B},\mathbb{P}_{X_t,X,Y,Z},\mathbb{D}_{X_t,X,Y,Z},\frac{1}{t^2}\Gamma_{X_t,X,Y,Z}\right),$$

converges D-narrowly to the product structure

$$\left(\mathbb{T}_d, \mathcal{B}(\mathbb{T}_d), \lambda_d, H^1(\mathbb{T}_d), G\right) \times \left(\mathbb{R}^{2d+q}, \mathcal{B}(\mathbb{R}^{2d+q}), \mathbb{P}_{X,Y,Z}, L^2, 0\right),$$

where G is the operator with constant coefficients

$$G[u] = \sum_{ij} \gamma_{ij} u'_i u'_j \quad \textit{with } \gamma_{ij} = \mathbb{E}\big[\Gamma[X_i, X_j]\big].$$

Proof. Let us limit the details to the fact concerning the convergence of Dirichlet forms.

If $F(a,b,c,d) \in \mathcal{C}^1 \cap \operatorname{Lip}\big(\mathbb{T}_\alpha \times \mathbb{R}^{2d+q}\big)$, then $\mathbb{E}\big[\frac{1}{t^2}\Gamma[F(X_t, X, Y, Z)]\big]$ has the same limit as

$$\mathbb{E}\Big[\sum_{ij} \frac{\partial F}{\partial a_{1i}}(X_t, X, Y, Z)\frac{\partial F}{\partial a_{1j}}(X_t, X, Y, Z)\Gamma[X_i, X_j]\Big].$$

However, according to the result on probability measures

$$\mathbb{E}\left[\frac{\partial F}{\partial a_{1i}}(X_t, X, Y, Z)\frac{\partial F}{\partial a_{1j}}(X_t, X, Y, Z)\Gamma[X_i, X_j]\right]$$

converges to

$$\mathbb{E}\big[\Gamma[X_i, X_j]\big]\int \frac{dF}{\partial a_{1i}}(a,b,c,d)\frac{\partial F}{\partial a_{1j}}(a,b,c,d)\,d\lambda_d(a)\,d\mathbb{P}_{X,Y,Z}(b,c,d). \qquad \diamond$$

Let us remark that the operator with constant coefficients G is obtained using the same formulae as that in the example following Definition VIII.1.

Example (Undamped coupled oscillators). Let us take the case of two identical oscillating (R, L, C) circuits with negligible resistance, coupled by a capacity, excited in the past by an electromotive force and now let out to free evolution.

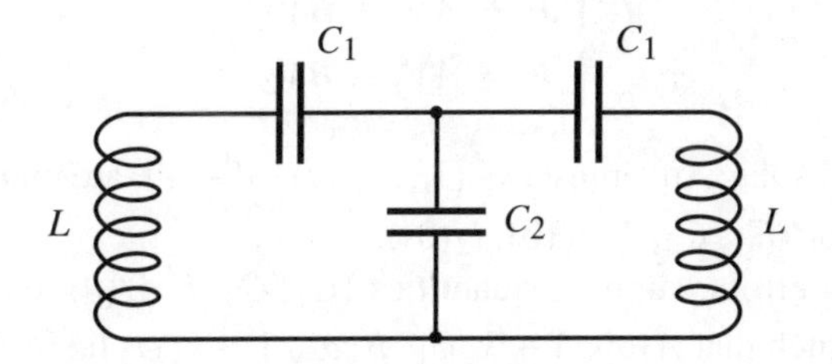

The intensities of the current are governed by the following equations

$$\begin{cases} L\dfrac{d^2i_1}{dt^2}+\dfrac{1}{C_1}i_1+\dfrac{1}{C_2}(i_1+i_2)=0 \\[2ex] L\dfrac{d^2i_2}{dt^2}+\dfrac{1}{C_1}i_2+\dfrac{1}{C_2}(i_1+i_2)=0. \end{cases}$$

setting $C=\frac{C_1C_2}{C_1+C_2}$, $k=\frac{C_1}{C_1+C_2}<1$ and $\omega=\frac{1}{\sqrt{LC}}$, we obtain

$$i_1(t)=\frac{C\omega}{2}\left[\frac{1}{\sqrt{1+k}}\cos(\omega t\sqrt{1+k}+\varphi_1)+\frac{1}{\sqrt{1-k}}\cos(\omega t\sqrt{1-k}+\varphi_2)\right]$$
$$i_2(t)=\frac{C\omega}{2}\left[\frac{1}{\sqrt{1+k}}\cos(\omega t\sqrt{1+k}+\varphi_1)-\frac{1}{\sqrt{1-k}}\cos(\omega t\sqrt{1-k}+\varphi_2)\right]$$

with

$$\cos\varphi_1=\frac{\sqrt{1+k}}{C\omega}\big(i_1(0)+i_2(0)\big)$$
$$\cos\varphi_2=\frac{\sqrt{1-k}}{C\omega}\big(i_1(0)-i_2(0)\big).$$

Let us assume the quantities C_1, C_2, L, $i_1(0)$ and $i_2(0)$, to be random and such that the pair

$$\big(\omega\sqrt{1+k},\omega\sqrt{1-k}\big)=\left(\sqrt{\frac{2C_1+C_2}{LC_1C_2}},\frac{1}{\sqrt{LC_1}}\right)$$

possesses a density with respect to the Lebesgue measure on $\mathbb{R}^2$.

Then, as $t\uparrow\infty$, the pair $\big(i_1(t),i_2(t)\big)$ converges in distribution to a pair of the form

$$(37)\qquad \begin{cases} J_1=A\cos 2\pi U_1+B\sin 2\pi U_2 \\ J_2=A\cos 2\pi U_1-B\sin 2\pi U_2 \end{cases}$$

where U_1 and U_2 are uniformly distributed on the interval $[0,1]$, mutually independent and independent of $(C_1,C_2,L,i_1(0),i_2(0))$. In other words

$$(38)\qquad \begin{cases} J_1=AV_1+BV_2 \\ J_2=AV_1-BV_2 \end{cases},$$

where V_1 and V_2 display Arcsinus laws $\big(1_{[-1,1]}(s)\frac{ds}{\pi\sqrt{1-s^2}}\big)$, are mutually independent and independent of $(C_1,C_2,L,i_1(0),i_2(0))$.

Regarding the errors, suppose quantities $\big(C_1,C_2,L,i_1(0),i_2(0)\big)$ are erroneous with hypotheses such that $A_1\omega\sqrt{1+k},\varphi_1,B,\omega\sqrt{1-k},\varphi_2$ lie in $\mathbb{D}\cap L^\infty$. For large t, the error on $\big(i_1(t),i_2(t)\big)$ can then be expressed by (37) with

$$
(39)\qquad \begin{cases}
\Gamma[U_1] = \dfrac{1}{4\pi^2}\mathbb{E}\left[\Gamma\left[\sqrt{\dfrac{2C_1+C_2}{LC_1C_2}}\right]\right] \\
\Gamma[U_2] = \dfrac{1}{4\pi^2}\mathbb{E}\left[\Gamma\left[\dfrac{1}{\sqrt{LC_1}}\right]\right] \\
\Gamma[U_1, U_2] = \dfrac{1}{4\pi^2}\mathbb{E}\left[\Gamma\left[\sqrt{\dfrac{2C_1+C_2}{LC_1C_2}}, \dfrac{1}{\sqrt{LC_1}}\right]\right] \\
\Gamma[U_1, A] = \Gamma[U_1, B] = 0 \\
\Gamma[U_2, A] = \Gamma[U_2, B] = 0
\end{cases}
$$

which yields

$$
\begin{aligned}
\Gamma[J_1] &= A^2\Gamma[\cos 2\pi U_1] + \cos^2 2\pi U_1\Gamma[A] + 2\cos 2\pi U_1 \cos 2\pi U_2\Gamma[A, B] \\
&\quad + B^2[\cos 2\pi U_2] + \cos^2 2\pi U_2\Gamma[B] \\
\Gamma[J_2] &= A^2\Gamma[\cos 2\pi U_1] + \cos^2 2\pi U_1\Gamma[A] - 2\cos 2\pi U_1 \cos 2\pi U_2\Gamma[A, B] \\
&\quad + B^2[\cos 2\pi U_2] + \cos^2 2\pi U_2\Gamma[B] \\
\Gamma[J_1, J_2] &= A^2\Gamma[\cos 2\pi U_1] + \cos^2 2\pi U_1\Gamma[A] \\
&\quad - B^2\Gamma[\cos 2\pi U_2] - \cos^2 2\pi U_2\Gamma[B].
\end{aligned}
$$

It should be understood that the renormalization property we applied (Proposition VIII.6) means that when t is large and as time goes by, the *standard deviation* $\sqrt{\Gamma[U_1]}$ of the error on U_1 *increases proportionally with time*. The same applies for the standard deviation $\sqrt{\Gamma[U_2]}$ of the error on U_2; the errors on A and B however remain fixed.

Concerning the errors, there is no, strictly speaking, stable asymptotic state: the asymptotic state is a situation in which the errors on U_1 and U_2 become dominant, they increase (in standard deviation) proportionally with time, and, in comparison, the other errors become negligible.

Example (Screen saver). A screen saver sweeps the screen of a computer in the following manner:

$$
\begin{aligned}
x(t) &= F\big(a + v_1 t\big) \\
y(t) &= F\big(b + v_2 t\big),
\end{aligned}
$$

where $F : \mathbb{R} \to [0, 1]$ is the Lipschitz periodic function defined by

$$
\begin{aligned}
F(x) &= \{x\} && \text{if } x \in [n, n+1[\\
&= 1 - \{x\} && \text{if } x \in [2n+1, 2n+2[, \quad n \in \mathbb{Z},
\end{aligned}
$$

$\{x\}$ being the fractional part of x,

$$
\{x\} = x - \max\{n \in \mathbb{Z},\ n \le x\}.
$$

Proposition VIII.6 states that if (a, b, v_1, v_2) are random and such that (v_1, v_2) have a density, $\big(x(t), y(t)\big)$ converges in distribution to the uniform law on the square $[0, 1]^2$ and moreover the error on $\big(x(t), y(t)\big)$ is asymptotically uncorrelated with those on (a, b, v_1, v_2).

Besides if screen-sweeping involves a rotation in addition to the translation

$$\begin{aligned} x(t) &= F\big(F(a + v_1 t) + F(\lambda \cos(\omega t + \varphi))\big) \\ y(t) &= F\big(F(b + v_2 t) + F(\lambda \sin(\omega t + \varphi))\big) \end{aligned}$$

the conclusion remains identical, since Proposition VIII.6 states that, once (v_1, v_2, ω) have a density, $\big(x(t), y(t)\big)$ converges to a pair of the form

$$\begin{aligned} X &= F\big(U_1 + F(\lambda \cos 2\pi U_3)\big) \\ Y &= F\big(U_2 + F(\lambda \cos 2\pi U_3)\big), \end{aligned}$$

where U_1, U_2, U_3 are i.i.d. uniformly distributed on $[0, 1]$ and independent of $(a, b, v_1, v_2, \omega, \varphi, \lambda)$. The law of the pair (X, Y), as easily seen, is the uniform law on the square $[0, 1]^2$.

Concerning the errors, by setting $v_3 = \frac{\omega}{2\pi}$, we obtain

$$\begin{aligned} \Gamma[U_i, U_j] &= \mathbb{E}\Gamma[v_i, v_j], \quad && i, j = 1, 2, 3, \\ \Gamma[U_i, \lambda] &= 0, && i = 1, 2, 3, \end{aligned}$$

and

$$\begin{aligned} \Gamma[X] &= \mathbb{E}\big[\Gamma[v_1]\big] + \lambda^2 \sin^2 2\pi U_3 \mathbb{E}[\Gamma[\omega]] \\ \Gamma[Y] &= \mathbb{E}\big[\Gamma[v_2]\big] + \lambda^2 \sin^2 2\pi U_3 \mathbb{E}[\Gamma[\omega]] \\ |\Gamma[X, Y]| &= \lambda^2 \sin^2 2\pi U_3 \mathbb{E}[\Gamma[\omega]]. \end{aligned}$$

Many other examples may be taken from mechanics for Hamiltonian or dissipative systems, or from electromagnetics, etc. where system dynamics eliminate both the probability measures and the error structures on the initial data and converge to typical situations. This argument of "arbitrary functions," i.e. probability measures with arbitrary densities always giving rise to the same limit probability law, therefore extends, up to renormalization, to error structures.

The limit structures obtained feature quadratic error operators with constant coefficients. This finding justifies, to some extent, that it is reasonable to take such an operator as an a priori hypothesis in many typical situations.

Nonetheless, the present section does not constitute the compulsory justification of the entire theory. It indeed proves interesting to found arguments in favor of one a priori error structure or another on model data. The study of error *propagation* through the model has however its own unique interest. This study gives rise to sensitivities and/or insensitivities of some of the results with respect to the hypotheses. As outlined in Bouleau [2001], the true link between error structures and experimental data is based on Fisher information and this is the subject of ongoing research.

Bibliography for Chapter VIII

Books

Ch. Berg and G. Forst, *Potential Theory on Locally Compact Abelian Groups*, Springer-Verlag, 1975.

N. Bouleau, *Processes stochastiques et applications*, Hermann, 2000.

N. Bouleau and F. Hirsch, *Dirichlet Forms and Analysis on Wiener Space*, Walter de Gruyter, 1991.

M. Brelot, *Eléments de théorie classique du potentiel*, Cours CDU 1959 Paris; New edition *Théorie classique du Potentiel*, Assoc. Laplace–Gauss, 1997.

B. Diu, C. Guthmann, D. Lederer and B. Roulet, *Physique statistique*, Hermann, 2001.

E. Engel, *A Road to Randomness in Physical Systems*, Lecture Notes in Statist. 71, Springer-Verlag, 1992.

H. Federer, *Geometric Measure Theory*, Springer-Verlag, 1959.

R. Kubo, N. Saitô and M. Toda, *Statistical Physics*, Vol. 1, Springer-Verlag, 1998.

R. Kubo, N. Hashitsume and M. Toda, *Statistical Physics*, Vol. 2, Springer-Verlag, 1998.

L. Landau and E. Lifchitz, *Physique Statistique*, MIR, 1967.

H. Poincaré, *Calcul des probabilités*, Paris, Gauthier-Villars, 2nd ed., 1912.

D. Tabor, *Gases, Liquids and Solids and Other States of Matter*, Cambridge University Press, 1991.

Articles

N. Bouleau, Calcul d'erreurs lipschitzien et formes de Dirichlet, *J. Math. Pures Appl.* **80** (9) (2001), 961–976.

N. Bouleau and Ch. Chorro, Error Structures and Parameter Estimation, *C.R. Acad. Sci. Paris Sér. I*, 2003.

L. A. Cauchy, Mémoire sur la rectification de courbes et la quadrature de surfaces courbes (1832), *Oeuvres complètes*, 1ère série, Vol. 2, 167–177.

J. Favard, Une définition de la longueur et de l'aire, *C.R. Acad. Sci. Paris* **194** (1932), 344–346.

E. Hopf, On causality, statistics and probability, *J. Math. Phys.* **13** (1934), 51–102.

L. S. Ornstein and G. E. Uhlenbeck, On the theory of the Brownian motion , *Physical Review* **36** (1930), 823–841.

H. Steinhaus, Length, shape and area, *Colloquium Mathematicum*, Vol. 3, Fasc. 1, 1954, 1–13.

M. E. Uhlenbeck and M. C. Wang, On the theory of the Brownian motion II, *Rev. Modern Phys.* **17**(2–3) (1945), 323–342.

Index